# Architectural Woodwork

# Architectural Woodwork

DETAILS FOR CONSTRUCTION

*Stephen P. Major*

JOHN WILEY & SONS, INC.
New York   Chichester   Weinheim   Brisbane   Singapore   Toronto

**A NOTE TO THE READER:**
This book has been electronically reproduced from digital information stored at John Wiley & Sons, Inc. We are pleased that the use of this new technology will enable us to keep works of enduring scholarly value in print as long as there is a reasonable demand for them. The content of this book is identical to previous printings.

This book is printed on acid-free paper. ☉

Copyright © 1995 by John Wiley & Sons, Inc. All rights reserved.

Published simultaneously in Canada.

No part of this publication may be reproduced, stored in a retrieval system or transmitted in any form or by any means, electronic, mechanical, photocopying, recording, scanning or otherwise, except as permitted under Sections 107 or 108 of the 1976 United States Copyright Act, without either the prior written permission of the Publisher, or authorization through payment of the appropriate per-copy fee to the Copyright Clearance Center, 222 Rosewood Drive, Danvers, MA 01923, (978) 750-8400, fax (978) 750-4744. Requests to the Publisher for permission should be addressed to the Permissions Department, John Wiley & Sons, Inc., 605 Third Avenue, New York, NY 10158-0012, (212) 850-6011, fax (212) 850-6008, E-Mail: PERMREQ@WILEY.COM.

This publication is designed to provide accurate and authoritative information in regard to the subject matter covered. It is sold with the understanding that the publisher is not engaged in rendering professional services. If professional advice or other expert assistance is required, the services of a competent professional person should be sought.

**Library of Congress Cataloging-in-Publication Data:**

Major, Stephen.
    Architectural woodwork: details for construction / by Stephen Major.
      p. cm.
    Includes index.
    ISBN 0-471-28551-X
    1. Architectural woodwork.    I. Title
TH1151.M35   1995                                       94-45678
694—dc20                                                      CIP

Printed in the United States of America

10 9 8 7 6

*To Ray Heath,
Craftsman and Friend*

# Contents

Preface    ix
About the Drawings    xi

## Section One    Background Information    1
Chapter 1. Materials    3
Chapter 2. Historical Precedents    13

## Section Two    Exterior Details    21
Chapter 3. Windows    23
Chapter 4. Doors    35
Chapter 5. Walls and Facades    47
Chapter 6. Roofline Trim    61
Chapter 7. Porches and Decks    71

## Section Three    Interior Details    85
Chapter 8. Windows    87
Chapter 9. Doors    93
Chapter 10. Ceilings    101
Chapter 11. Walls    109
Chapter 12. Floors    113
Chapter 13. Fireplace Woodwork    125
Chapter 14. Cabinetry and Shelving    133
Chapter 15. Stairs and Balustrades    143

## Section Four    Special Topics    153
Chapter 16. Curves    155
Chapter 17. Exterior Protection    167

Glossary    177
Appendix 1: Wood Mouldings    191
Appendix 2: Wood Siding    197
Appendix 3: Wood Grades    201
Appendix 4: Plywood    205
Appendix 5: Cedar Shakes and Shingles    213

Index    215

# Preface

"WOODWORKING IS A LOST ART!" This is a common contention of many individuals who cherish the impossibly crafted furnishings of past centuries. It's one more variation on the lament that "They just don't do things the way they used to." My answer to the first statement is that it is a half-truth; the craft and art of using and working wood is very much alive and well today—it simply doesn't resemble the discipline taught our forebears. To the second claim I can only answer, "Very true—and how very, very fortunate!"

Modern attitudes, priorities, and technologies have enabled wood to be used on a grand scale to serve the construction needs of a burgeoning population. It may not always be in its traditional form, and we may not work it by tedious, backbreaking methods, but it is wood nonetheless, and its uses must be designed and specced with as much care as ever. And when wood no longer serves a purpose well, we can now deftly substitute an alternate material that is functionally or financially superior.

To design and spec constructions that use wood or wood substitutes as functional or finish materials requires a firm knowledge of the applications and limits of the materials, as well as an understanding of the practical and aesthetic significance of wood construction—its background and its modern interpretations.

*Architectural Woodwork: Details for Construction* is a reference and design guide for architects, designers, builders, building owners, and skilled tradespeople who are responsible for the selection, design, specific detailing, and installation of architectural woodwork.

Most nonstructural woodwork that is part of a building (as opposed to furniture and sculpture) is defined as architectural woodwork. Additionally, any building material that was traditionally made of wood but is now manufactured of an alternative material and used as architectural woodwork remains within the scope of this book. Hence, aluminum millwork, vinyl siding, and styrene reproduction mouldings are addressed alongside their traditional wood counterparts.

The broad range of architectural woodwork is covered herein in a logical organizational system. Following two general chapters on materials and historical precedents, the book is divided into two main sections: exterior details and interior details. Items that transcend this division (windows,

## PREFACE

doors, and walls, for example, have both inside and outside exposures) are treated appropriately in both sections.

Each section is composed of several chapters describing specific parts, surfaces, or areas. The section on exterior details is divided into chapters on windows, doors, walls and facades, roofline trim, and porches and decks. The section on interior details has chapters on windows, doors, ceilings, walls, floors, stairs and balustrades, fireplace trim, and cabinetry and shelving. A third section has chapters on the special problems of curves and exterior protection.

Each subject is treated in detail, with written descriptions of practical background, design, and installation considerations, followed by illustrations of pertinent construction details.

In an effort to keep things simple, virtually every illustration is presented without dimension, in order that the details may be adapted in whole or in part to any suitable application. All dimensions that are given in the text are in inches. The common trade names of woods are used throughout the text; their specific names (and properties) are presented once, in Chapter 1.

It is my sincere hope that the material presented in this book will be of real value to those who find conflict and confusion in the current supply of information on this vast subject. As a professional builder, I am allowed, almost daily, the privilege of working with the concepts, designs, and construction materials that form the basis of this book. I see what works and what fails, and I usually get to find out why in a hurry. The materials presented herein are *basic* concepts and guidelines—the information you need *before* you tackle the manufacturer's literature (which is always slanted), and *before* you struggle to devise a solution to an old problem (it's likely been done already).

I am grateful to several people who have lent their help, in one way or another, to getting this book into printable form: Richard Elbert, for his early encouragement; David Wenz and my father, Joseph, for taking up the slack during my frequent absences from job sites; Nina Dugan, who lettered each illustration; and especially my wife, Cherie, who continually strived to keep this effort at the top of our priority list.

# About the Drawings

MOST OF THE ARTWORK presented in this book was hand-drafted and inked by the author. After one xerographic reduction, the lettering was added, again by hand.

There are several types of drawings used. The most common is the sectional view, or cross-section, which illustrates how a particular construction would appear if it were sliced through. An example of a cross-section is Figure 6-7, which shows particular roofline construction details. Cross-sections illustrate important construction details and relationships.

Some of the illustrations are elevation views. The elevation is a fairly realistic rendering of the vertical surface of an object, viewed head on, without perspective distortion. Examples of elevations are Figures 2-2 and 2-3. Elevations provide useful approximations of outward appearance.

Some of the illustrations are what I call "extended cross-sections." These information-packed drawings are a combination of cross-section and elevation views. They are full cross-section drawings imposed upon the elevation views they depict. An example is Figure 10-4(a), which portrays several crown moulding constructions. Extended cross-sections are useful because they illustrate the direct relationship between a construction (shown in the cross-section) and its appearance (shown in the elevation).

A few of the drawings are of the pictorial type: perspective, isometric, or orthographic. These show an object in three dimensions. Perspective (Figure 2-8) is the most realistic, as it incorporates one or two "vanishing points" to depict diminishing size with increasing distance. Isometric (Figure 1-1) and orthographic (Figure 7-6) do not.

In order that each convey its point in a clear manner, certain drawings are not full detail renderings as would be used on a set of plans. For example, exterior details (especially sections) may show simplified or even nonexistent framing/interior details; Figure 5-6 illustrates exterior corner details, yet shows minimal framing or interior components, since these may vary considerably and do not add substantively to the points being presented.

# Architectural Woodwork

# Section One
## Background Information

CHAPTER 1 | **Materials**

Of all the materials used to fabricate architectural woodwork—woods, metals, plastics, and various adhesives—the most common material is, of course, wood. Fifty years ago, virtually all of the constructions discussed in this book were made of wood and nothing else. And while wood is still widely used for the same applications now as it was then, many alternate materials have found a place as substitutes for wood in architectural woodwork. Some of these materials are equal to the real thing, some are superior, and some are a sad compromise.

Wood is available in seemingly countless forms meant to serve numerous ends. Hundreds of species of trees provide solid wood; dozens of them are commercially significant. Likewise, correspondingly numerous veneers are used for the face and core of plywood, one of the most widely used wood products and a significant material in architectural woodwork.

Beyond the familiar grained face of solid wood and wood veneer, wood in chip, particle, or fiber form bonded with adhesives constitutes a growing percentage of the materials used in woodwork, either as core or base substances (in faced sheet goods) or as stand-alone materials (in medium density fiberboard and hardboard). Each of these materials has a place, and each is suited to particular uses.

Metals, widely used as a direct substitute for wood in architectural woodwork (e.g., on windows) or as an accessory material (e.g., as flashing), are available in many forms to suit various uses. Steel is used for doors, frames, windows, flashings, and other accessories, and is available painted, coated, galvanized, plated, and in stainless (noncorrosive) form. Aluminum is used for similar applications; it is available in sheet or extruded forms, either with a mill finish or anodized or coated. To a lesser extent copper, lead, and zinc, largely in sheet form, have a place in certain constructions as durable flashings and gutters. And most fasteners used in the manufacture and installation of architectural woodwork (nails, screws, cleats, pins) are made of various metals.

Plastic is increasingly used as a wood substitute. *Plastic* is a generic term—plastics appear in an endless variety of common and uncommon forms. Moulded vinyl plastics are used as siding and for exterior trim, millwork cladding, and even entire window frameworks. Rigid foam plastics (e.g., styrene) are available in nearly any moulded shape imaginable, including entire architectural structures, and in many styles.

Lastly, a seemingly minor yet significant class of materials used in architectural woodwork is adhesives and sealants. Many chemical formulations are used to bond various substances, both in the manufacture of specific materials (like plywood) and in the installation of woodwork components. Sealants are widely used as barriers to infiltration and to facilitate the application of a good finish. A basic understanding of their applications and limits is useful, as the inadequate or incorrect use of adhesives and sealants may result in unsuccessful woodwork.

## Wood

### Structure and Properties

Although wood is one of the most familiar, versatile, and useful materials on earth, its quirks and idiosyncrasies are unfamiliar to many who use it or specify it.

As a tree grows it produces cells, generally tubular and elongate in shape, in a living layer known as the cambium, which exists just below the bark. New cells press against older ones, and these older cells, nearer the center of the tree, gradually die and harden to become solid wood.

Since growing conditions are generally best in the spring, the cells produced then are large and carry volumes of water. The contrast between these cells and the smaller summer-growth cells causes the formation of the familiar growth rings, by which one can estimate the age of a tree. These rings, when exposed longitudinally in the making of boards, cause the grain patterns that visually distinguish wood species. Figures 1-1 and 1-2 illustrate these properties.

It is the wood cells—their shape, size, orientation, and interconnection—that give wood its important and consequential characteristics. Wood is essentially an agglomeration of tightly packed, vertically oriented tubes (except the ray cells, which are oriented radially), each overlapping and bound to the next to form a structure of surprising strength and resilience. Like any cylinder, a wood cell is somewhat resistant to bending and to shear forces perpendicular to its length, and it is highly resistant to compressive or tensile forces parallel to its length. With its hundreds of millions of essentially identical bound and interlocked cylinders, wood displays unmatched structural integrity, having a strength-to-weight ratio higher than most other materials.

Fortunately, despite its great strength, wood has excellent workability. It can be easily cut, shaped, bent, steamed, sliced, and split. And after all that, it can be glued, pressed, nailed, screwed, preserved, and painted. The craft of working wood into usable forms has developed continually over the centuries, and today we are at the point of utilizing wood to possibly its fullest extent ever.

On the down side, wood possesses two negative characteristics that greatly affect its success as a building material. First, it burns, and burns well. All those cells are tiny storehouses of concentrated carbohydrate energy, and when this energy is released by rapid oxidation (burning) the cells quickly self-destruct with a furious exothermic reaction. This characteristic of wood carries grave consequences, and countless lives have been lost throughout history in burning wooden buildings.

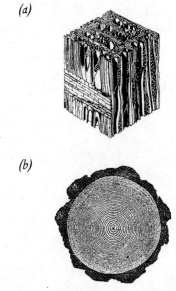

**Figure 1-1**
*The cellular structure and growth rings of wood (courtesy USDA Forest Products Lab).*

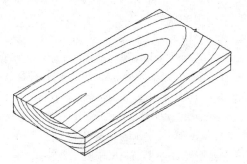 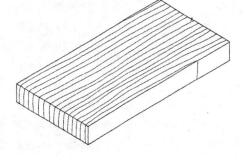

**Figure 1-2**
*Longitudinal cutting along growth rings reveals wood's distinguishing figure, or grain.*

Second, wood cells, although they're dead, will forever continue to attract water and will shrink and swell in response to their water content. Water does not need to be in liquid form to affect wood; water vapor causes dimensional change in wood as well. The process proceeds by simple diffusion along a concentration gradient. Even multiple layers of water-resistant finishes do not completely eliminate this continuous process.

An additional consideration that affects wood's design utility is interspecies variation. All woods are not created equal, and red oak, for example, is suitable for quite a different range of uses than red cedar. Specific gravity, dimensional stability, porosity, decay resistance, finish-holding ability, and several other characteristics will vary with the species and even according to the particular conditions under which a specific tree grew.

Each of wood's properties—its strength, workability, flammability, and affinity for water—as well as its species-related properties, must be considered when using wood as a construction material.

## Types of Wood

Based on fundamental botanical differences, trees and their wood are classically divided into two major categories: the softwoods (coniferous, needle-bearing, with naked seeds), and the hardwoods (deciduous, leaf-bearing, with covered seeds). Generally speaking, hardwoods are harder than softwoods, although there are common instances of the opposite condition.

Table 1-1 provides a general classification of woods in common commercial use today. In a broad sense, common softwoods are used for structural purposes in light wood framing, and common hardwoods are used for decorative or finish constructions. Natural exceptions to this simplification exist, of course; pine, cedar, and redwood, all softwoods, are widely used for finish-grade millwork; oak, a hardwood, is used as a structural material in timber-frame construction; and poplar, a bland hardwood, is used in the manufacture of oriented strand board, a structural panel.

Aside from a consideration of its suitability for a particular use and its distinct properties, the selection of a wood type is typically based on additional-factors such as availability, local practice or tradition, taste and style, and comparative cost.

## Lumber

Wood has limited use in log form and must be processed into wood products before it is suitable for an end use such as architectural woodwork. Once a tree is felled, it is often allowed to season for a time (to lose some of its free water) before it is sawn into boards. Logs are typically sawn with an intended purpose in mind, for optimum yield. A variety of approaches can be used, but for the sake of simplicity, the resultant lumber can be classified as either plain-sawn (tangentially cut) or quarter-sawn (radially cut), as shown in Figure 1-3. The orientation of the cut relative to the direction of the growth rings determines a variety of lumber characteristics, such as dimensional stability and grain figure. Since wood shrinkage is greatest along the direction of a tree's growth rings, a tangentially cut board will shrink (and likely warp) more than a radially cut board from the same tree. Quarter-sawn hardwood and so-called vertical-grained softwood are highly valued for their stability.

## Table 1-1 Selected Characteristics of Common Woods[1]

| Commercial name | Genus species | Specific gravity (density)[2] | Resistance to decay[3] | Shrinkage[4] | Relevant uses[5] |
|---|---|---|---|---|---|
| *Softwoods* | | | | | |
| Cedar, northern white | Thuja occidentalis | .32 | high | low | Poles, lumber, fencing, shingles |
| Cedar, western red | Thuja plicata | .32 | high | low | Siding, trim, shingles, posts |
| Cypress, bald | Taxodium distichum | .46 | high | moderate | Lumber, siding, flooring, millwork |
| Douglas fir, west | Pseudotsuga Douglasii | .50 | moderate | moderate | Doors, millwork, flooring, plywood, structural lumber |
| Hemlock, western | Tsuga heterophylla | .45 | low | high | Lumber, flooring |
| Pine, southern yellow (loblolly) | Pinus taeda | .51 | low | high | Structural, millwork, flooring, plywood, extensive outdoor use as pressure-treated lumber |
| Pine, ponderosa | Pinus ponderosa | .40 | low | moderate | Window and door parts, interior and exterior trim |
| Pine, western white | Pinus monticola | .38 | low | moderate | Misc. millwork, lumber, trim, siding |
| Redwood, heart | Sequoia sempervirens | .40 | high | low | Window and door parts, misc. millwork, decking, structural lumber |
| Spruce, white | Picea glauca | .40 | low | moderate | Structural lumber, secondary wood in cabinetry, millwork |

## Table 1-1 (continued)

*Hardwoods*

| | | | | | |
|---|---|---|---|---|---|
| Ash, white | Fraxinus americana | .60 | low | high | Flooring, cabinetry, veneer |
| Birch, yellow | Betula lutea | .64 | low | high | Plywood, cabinetry, mouldings, flooring |
| Maple, sugar | Acer saccharum | .63 | high | high | Flooring, veneer, cabinetry |
| Oak, red | Quercus rubra | .63 | high | high | Flooring, veneer, interior trim, cabinetry |
| Oak, white | Quercus alba | .68 | high | high | Flooring, misc. trim, structural timbers, posts |

1. From U.S. Forest Products Laboratory
2. Values are for dry wood
3. Heartwood only
4. Volumetric shrinkage from green to 0% moisture: low (<8%), moderate (8–12%), high (>12%)
5. Uses that apply to this text

Green, rough-sawn boards are typically dried in a kiln under controlled conditions to drive enough moisture from the wood to allow its use as a predictable material. Depending on its intended destination, the lumber is then graded according to its dimensions and the quantity and type of its defects. Further conversion, in the form of dimensioning, planing, resawing, or other milling and processing, is undertaken by intermediate handlers or end users.

Lumber must be utilized in accordance with its unique physical properties. The successful use of solid wood in constructions of close tolerance and high value depends on an understanding of construction principles that

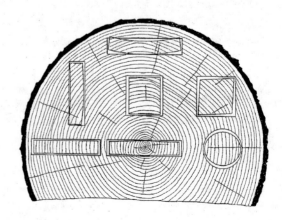

**Figure 1-3**

*Different approaches to sawing a log into boards yield a variety of grain patterns and consequential lumber properties. Quartersawing (left) cuts growth rings straight across their short dimension, minimizing dimensional instability (courtesy USDA Forest Products Lab).*

respect the hygroscopic (water-attracting) nature of the material and an ability to apply these principles in a skillful manner.

## Plywood, Veneer, and Other Panels

Solid wood, with its peculiar properties, requires a certain amount of care to manufacture into usable products. It must be treated and handled painstakingly in order to attain and then maintain its ideal moisture content and its qualities of dimension and shape. It must be worked by skilled individuals who have a thorough understanding of its properties and their consequences. And as it is an organic material, its quality, form, and character are highly variable; it must be graded and sorted to retain the particular characteristics required for a specific use. In other words, it does not easily fit into the dollar-driven process of mass production and mass distribution. Nevertheless, manufacturers have found ways to preserve wood's desirable qualities while minimizing its negative effects on efficient manufacture.

One of the most common wood-based materials is the man-made panel, or sheet good. These materials are manufactured in a variety of configurations, but virtually all of them are laminar in their structure, composed of layers of wood or layers of wood particles at various densities or orientations. Laminar sheet goods usurp the two-way structure of solid wood; they are much more homogenous and therefore exhibit much greater dimensional stability and predictability than solid wood.

Plywood, which is composed of a core between faces of wood veneer (thin slices of wood), is a widely used substitute for solid wood. Its veneered faces maintain the appearance and feel of solid wood, and its core, engineered to be much more dimensionally stable than solid wood, offers superior accuracy, integrity unaffected by grain and growth factors, and predictability. The core may contain layers of additional veneers laid at right angles to the face veneers and to each other (veneer core), narrow strips of solid wood in series (solid core), or a monolithic composite material such as particleboard or fiberboard. As the edges of all plywoods (except solid core plywood and certain densely laminated veneer core sheets) are unattractive if exposed, it is common practice to conceal the edges with various edgings or bandings, depending on the intended use. Sheet goods are also available with nonwood faces; various plastics are commonly laminated to particleboard for use as case panels.

The wood veneers used to form the face and core of plywood are generally selected for their similarity in dimensional stability, to minimize warpage. Veneers are typically peeled from the log (rotary sliced) in large, continuous sheets. Face veneers of plywood chosen for appearance are sliced rather than peeled, with the result preserving the appearance of true boards.

These necessarily narrow-sliced veneers are visually matched for uniform or decorative appearance and carefully joined at their edges into a single sheet large enough to cover the entire plywood panel. Just like solid lumber, the veneered faces of plywood are graded by quality.

Monolithic sheet goods are composed of wood fibers, particles, or wafers pressed under heat and pressure, often with large volumes of adhesives to form stable panels. Fiberboard and hardboard have uses as paint-grade woodwork; hardboard also comes in the form of linear siding materials, often with water repellants and preservatives added. Particleboard, available in several densities, is used as underlayment, core stock, and sheathing. Oriented strand board, composed of wood wafers roughly arranged in layers of opposing grain direction, is commonly used as a structural sheathing panel.

## Nonwood Materials

### Metals

Metals used in architectural woodwork are of two types: direct wood substitutes, such as doors and windows, and accessory materials, such as flashing and hardware.

Steel, generally in sheet form of various gauges, is widely used as door bucks (frames), door panels, window sashes and frames, and a variety of functional and decorative hardware. Steel is protected from corrosion by applying various coatings (paint, powder coat, galvanized zinc), or by alloying it with other metals (nickel and chromium) to reduce its reactivity. These are known as stainless steels.

Zinc and zinc alloys, in sheet form, make malleable, corrosion-resistant flashings.

Aluminum, in sheet or extruded form, is commonly used much as steel. It is naturally resistant to corrosion; it is commonly coated or finished more for appearance than for protection, by anodizing or enameling. Steel and aluminum in sheet form are brake- or roll-formed into flashing, trim, roofing, and decorative claddings on exterior surfaces. Aluminum is also extruded into a variety of trims, mouldings, channels, and window frames.

Copper, in sheet form, is available in several gauges and degrees of hardness, for use as highly durable roofing, flashing, gutters, downspouts, and screening. It is available with a lead coating to eliminate its tendency to develop a green patina (particularly in industrial atmospheres). Copper alloyed with varying proportions of zinc produces brass and bronze, widely used in cast and forged configurations as hardware and fasteners.

## Plastics
Like metals, plastics used in architectural woodwork can take the form of direct wood substitutes or accessories.

Extruded vinyl plastics are widely used for exterior trim and siding and for window cladding and window frames and sashes, particularly in residential applications. Polystyrene and polyurethane plastics are used to form moulded architectural details such as cornice mouldings, casings, and columns.

High-pressure plastic laminates (layered kraft paper and melamine plastic bonded with phenolic resin) are highly resistant to water, alcohol, and stains, and are widely used as an overlay on a base of core stock for countertops, tabletops, and cabinet components.

Reinforced plastic fiberglass is used to form door panels that have the appearance of real wood.

## Adhesives, Caulks, Sealants
Wood-bonding adhesives commonly used today are synthetic resins, as opposed to the older animal- and vegetable-based glues. Phenol formaldehyde, resorcinol formaldehyde, and melamine formaldehyde form bonds that are completely waterproof; these are commonly used to join exterior-grade plywood veneers and other exposed constructions. Urea formaldehyde forms water-resistant joints, and like resorcinol it can be formulated to bond at room temperature. The other resins require the application of heat to cure. Polyvinyl acetate (PVA) emulsion, commonly known as standard wood glue, is generally used for woodwork and furniture assembly where maximum strength and water resistance are not critical.

Caulk, originally a mixture of oils, fillers, and various working agents, is commonly used as a joint filler when minimal movement between building components is expected.

If greater movement is expected—usually due to thermal expansion and contraction—then sealants (rubbery elastomeric materials) are often used to seal spaces and fill crevices. Many formulations exist, including one- and two-part compounds of acrylics, butyls, silicones, and polyurethanes. Sealants should be carefully chosen for appropriate properties in relation to the substrates being joined.

CHAPTER 2 | # Historical Precedents

The forms seen in modern architectural woodwork—the profiles of mouldings; the arrangements of window and door trims; the constructions of doors, walls, and cornices—are each the product of centuries of technical and stylistic evolution.

The ancient builders of long-vanished wooden temples (later replicated in stone) set certain standards in their methods of addressing basic building problems like defining space and deflecting the elements, within the practical constraints of their tools and materials. Much Western architecture has its roots in the elemental buildings that were the precursors to the inspirational classical structures still standing today, and with good reason—these classical structures, however ornate, were elegant in their simplicity of order and form.

Each period of history has been marked by the architectural styles of the day—they have defined history as much as history has defined them. Many timeless styles have been repeated in various configurations over the centuries, and many of today's architectural forms and details are reproductions of successful styles developed long ago.

Since the long history of the development of architectural forms is beyond the scope of this book—and is the subject of many other volumes—this chapter instead provides a limited analysis of some basic concepts central to the development of all forms of architectural woodwork. Elemental building concepts are presented to illustrate their relationship to modern woodwork forms; field and edge analyses, and their implications for woodwork design, are introduced; and a comparison is made between traditional and contemporary woodwork forms.

# CHAPTER 2

## Elemental Building Concepts

In the simplest (classical) sense, a wooden building is an articulation of several basic pieces, each configured to support and define the whole. The first element is the vertical column, post, or stud, across which is placed the second element: the horizontal beam, or lintel. Ceiling joists are laid horizontally atop the beams, and canted rafters (to shed snow and rain) are placed to define the roof. Figure 2-1 illustrates these elemental parts. Finally, the boards and panels that form the walls, roof, doors, and partitions are inserted to fill and enclose the basic framework. This type of elemental structure acknowledges the laws of gravity, and thus—as materials are laid upon each other—most joints are horizontal.

Early versions of such elemental structures were largely made of logs; squared timbers and eventually flat boards came into use as technology advanced. With these advances came order, and early builders created successful designs that became the inspiration for later work. The use of the round column, thickening toward the base, became conventional, both for its deference to the form of the tree and for its apparent and real stability. Rectangular plans were a natural companion to the use of straight timber. Canted rafters, with overhanging ends to divert water from the vulnerable (often clay) walls below, became customary for practical reasons.

Over time, ancient architects refined the functional and aesthetic attributes of the elemental parts of buildings and, more importantly, developed sophisticated orders of form to deal with the crucial intersections of each basic component. These intersections and their attendant coverings, mouldings, and embellishments make up much of the woodwork presented in this book. The historical significance of intersecting elements is further explored in the following section on fields and edges.

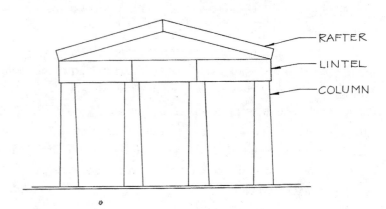

**Figure 2-1**
*Elemental building components.*

# Historical Precedents

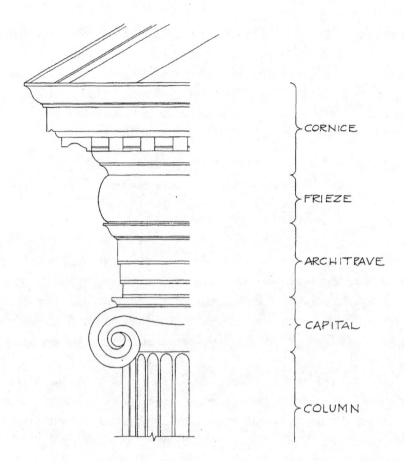

**Figure 2-2**
*The basic elements of a classical order (the Greek Ionic). The entire structure between the column capital and the roof edge is the entablature.*

A column's intersection at its base with a pedestal or with the earth required a moulded form, as did its top. Entire sequences of moulded forms, each with a distinct aesthetic and practical purpose, were developed to define the space from the top of the column (the capital) to the roof dripline. These "entablatures," together with the form of the column itself, became the classical "orders," from which many styles of construction and decoration have sprung, mimicking them exactly or, at least, in their gross form. The orders were applied universally, in three dimensions as exterior superstructures and in lower relief as the boundaries of openings or niches. Figure 2-2 illustrates a classical order (the Greek Ionic), showing the basic building elements and the mouldings used to cover and embellish the joints between them. Architectural historians have recognized that many seemingly decorative moulded forms have their basis in utility. For instance, the cyma, or crown, of the cornice is derived from the curving form of a wooden or clay rain gutter, and repeating dentils mimic the exposed ends of overhanging rafters or projecting joists.

No wooden classical elemental buildings remain, but highly refined stone versions, largely ancient temples in Greece and Rome, have thankfully fared much better. It is these stone structures, wholly based on their short-lived wood precursors, that remained to inspire and reinspire the work of generations of builders and architects. The classical forms—the moulded profiles, the characteristic proportions, the sequences of the orders—have been used, adapted, and sometimes misused throughout Western architectural history. Even as advancing technology conspires to make these forms ever more obsolete, they remain all the more useful for their simple functional grace.

## Fields and Edges

In order to better understand the practical implications of the design and use of decorative architectural woodwork, it can be helpful to apply a simple scheme of organization to the various parts of a building. This scheme need not be cumbersome or unwieldy—in fact, to meet the needs of the brief discussions in the following chapters, only two classifications are necessary: fields and edges.

Each discrete surface or major component of a building can be considered a *field*. Walls, ceilings, floors, roofs, windows, and doors are all fields. They are large, obvious, and homogeneous, and they serve primary functions: as basic construction elements, boundaries of spaces, and functional portals.

Where one field meets another, field conditions change. The line at which field conditions change can be considered an *edge*. Edges of fields are the zones of transition from one field to another. Edges are generally smaller and more subtle than fields, and they serve secondary functions. Figure 2-3 illustrates some general examples of the relationships between fields and edges, and Table 2-1 summarizes their properties.

Edges in architectural woodwork can range from minimal, as in a drywall corner; through simplistic, as in a flat casing; through complex, as in a wide cornice; to elaborate, as in the capital of a Corinthian column. Fields (walls, floors, and so on) tend to be plain, yet they may also sport embellishment to varying degrees—siding, parquetry, appliqué. Fields are subject to a different set of design constraints or considerations than edges, because they are usually large and dominant. A small change in the specification of a field can cause a large change in overall appearance, function, and cost.

A useful, if punning, analogy to aid in understanding the significance of edge conditions is to equate the problem of building design with the task of mowing a lawn. To successfully accomplish this seemingly mundane chore, the mower must first mindlessly traipse back and forth across the lawn, clipping

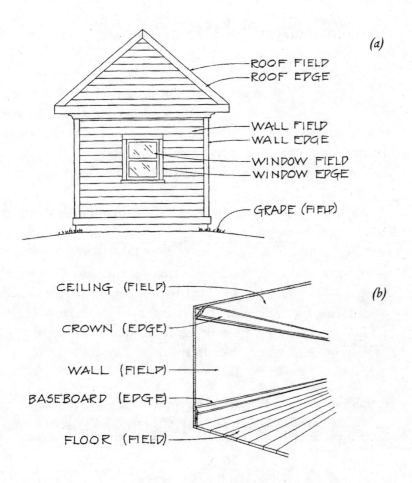

**Figure 2-3**
*Fields and edges in common construction. (a) Exterior (b) Interior.*

the grass in a long series of monotonous, repetitive runs, perhaps on a riding mower or tractor if the lawn is large. This portion of the work is the field. It is the bulk of the lawn, and any one part of it is very much like any other.

## Table 2-1 Properties of Fields and Edges

| Attribute | Fields | Edges |
| --- | --- | --- |
| Relative size | larger | smaller |
| Function | primary: fundamental components | secondary: delineation of and transition between fields |
| Discernability | visible at long range | visible at close range |
| Prominence | conspicuous | subtle |
| Examples | walls, floors, roofs | corners, trims, frontiers |

The remainder of the work—trimming the edges around trees, garden borders, and so on—requires a bit more skill and a little more attention to detail. If it's done sloppily, it will cause the whole job to look bad. Such is the case with architectural woodwork. It is a relatively simple task to design a space with four walls, a floor, a ceiling, windows, and a door. But it is quite another task to connect each of these discrete elements with appropriate, subtle transitions. This is the challenge in designing edge details—this is where one must pull out the small hand mower, the clippers, and the edger, and go to work creating something special.

The history of architectural woodwork has consisted, in large part, of the development of methods of dealing with edges. The trims, structures, and mouldings that form the transitions between fields have afforded building designers with zones of opportunity—areas for embellishment and creative expression—which are often so distinctive that they alone define a building's style. Classical orders are really only stacks of fields (column, beam, roof) joined with specific mouldings or treatments that are, in fact, edges. The fields *form* the building, but the edges *make* it, so to speak.

## Traditional and Contemporary Details

Taking the design attributes of the classical orders as a basis, it can be argued that the essence of traditional design in architectural woodwork has been the decoration of bland fields with contrasting and obvious edges. An example is shown in Figure 2-4. Depending on the designer's intent, edges may be embellished to the point of totally dominating the field, even to the point of appearing excessive or ostentatious (see Figure 2-5).

A good traditional edge design provides a gradual, graceful, and well-proportioned transition between fields. This is illustrated conceptually in Figure 2-6. Conversely, a poor edge design lacks grace and is too abrupt or inharmonious to form a gradual or smooth transition between fields. Figure 2-7 compares "good" edge conditions with "poor" ones, using various crown and base mouldings as examples.

Such rules are largely ignored in contemporary design, and the emotions evoked by a contemporary backdrop are quite different from those provided by the familiarity of traditional style. With its base in modern materials

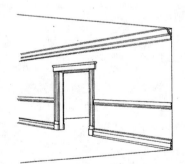

**Figure 2-4**
*A traditional approach to edge conditions often employs prominent moulded forms.*

**Figure 2-5**
*Overbearing edges.*

and methods, and its execution often guided as much by budgets and efficiency as by a quest for a definitive order, the essence of contemporary design is to place fields in stark, sudden adjacency. Walls turn into windows and ceilings with a sharp edge (see Figure 2-8); there is no transition, just a clean line. Gradual grace is often too rich, too elaborate, and too expensive.

**Figure 2-6**
*A "good" conceptual edge (left). The reversal of flow (right) is less successful.*

**Figure 2-7**
*Edges. (a) Although these edges are similar, the crown profile at right breaks the smooth flow of the (better) assembly at left. (b) The ability of these baseboards to form a gradual transition between wall and floor diminishes from left to right.*

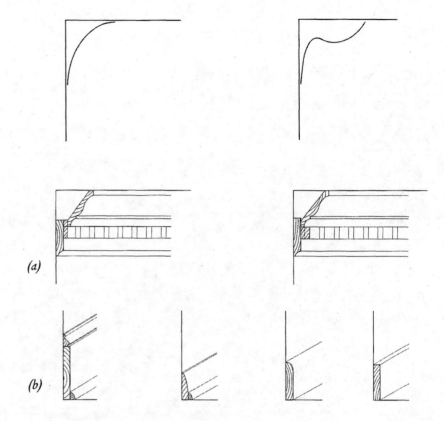

Since field conditions are often identical in both contemporary and traditional schemes (smooth walls, plain ceilings) it is often simply the differences in edge conditions that distinguish each school as unique. Neither is more or less correct than the other, and if the ancients had had today's technology and priorities, the classics might bear a striking resemblance to modern contemporary design: basic building elements (fields) joined at their junctures with functional and cost-effective transitions (edges).

**Figure 2-8**
*A typical contemporary approach tends to use minimal, linear edges.*

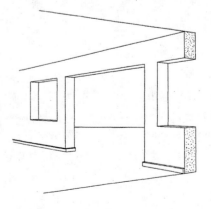

# Section Two
## Exterior Details

CHAPTER 3 ‖ Windows

Since its ancient beginnings as oiled paper fastened over a crude hole, the window has been an important and powerful architectural element.

A window can provide light to illuminate interior spaces. It can offer a view of the outdoors to the inhabitants of a dwelling. It can provide the simplest form of ventilation—an open window offers free air exchange between a building's interior and exterior environments. Judiciously placed, a window can handily add heat to a building by admitting the sun's radiation. A window brings the outdoors in.

Beyond fulfilling these basic, practical functions, windows and window arrangements have the potential to make powerful and beautiful design statements. Windows may be placed singly or arranged horizontally in rows, vertically in stacks, or en masse in banks. They can be made into nearly any shape. They can be trimmed with a full complement of traditional mouldings, or they can appear as stark blocks of light punched through a contemporary wall.

Because windows can possess so many different functional and aesthetic attributes, and since their effect on a building's performance and appearance can be profound, a designer must approach the selection and use of windows with the utmost care and attention.

This chapter explores and presents window design and detailing in the context of window field and edge conditions (see Chapter 2). The dominant field material for windows is glass, commonly mounted in a frame. Both glass and frames are available in many varieties. Window glass may be tinted, coated, laminated, doubled, or tripled. Window frames may be of wood, plastic, metal, or a combination of materials. A window can be operating or fixed. The sashes of operating windows can slide, swing, tilt, or pivot.

# CHAPTER 3

The edges of windows, and the juxtaposition of window fields, provide nearly unlimited opportunities for creative and stylistic expression. A range of exterior trim alternatives is presented herein, from stark to complex, along with pertinent construction details for installation, sealing, and flashing.

## Window Construction

The construction of two typical windows is illustrated in Figure 3-1. Figure 3-1(a) is a diagram of a typical modern double-hung wooden residential window, with its exterior exposures sheathed in protective aluminum cladding. The important parts are labeled. Figure 3-1(b) shows a typical modern commercial fixed window, of aluminum with a thermal break. Again, the important parts are labeled.

**Figure 3-1**
*(a) A clad residential window. (b) A metal commercial window.*

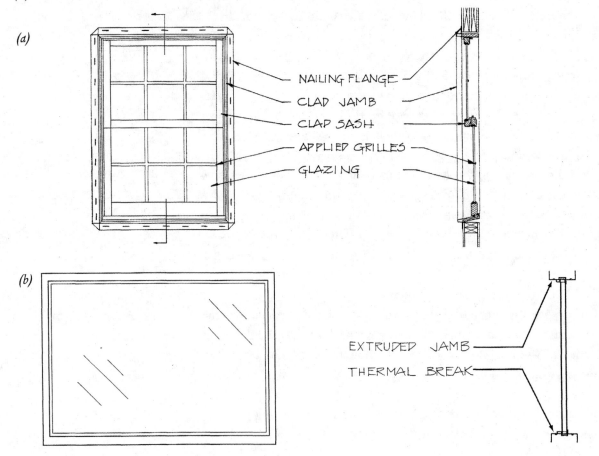

## Residential versus Commercial Windows

Windows are marketed today in two major types: the commercial window and the residential window. There are some basic differences in their construction, since their end uses are somewhat different. Residential windows are, however, often used for light-duty commercial service, and commercial windows do see some use in certain residential applications.

Although all reputable windows are durable and well built, windows designed for commercial use are generally built with several specific additional features in mind. They are typically constructed to withstand long, severe service with little or no maintenance; their initial cost is not nearly as important as their life-cycle cost. They are typically built in a plain fashion, with clean, narrow lines, in order to meld with a range of building styles without conflicting with other exterior or interior elements of a building's appearance. Commercial windows are usually available in any specified size, made-to-order. If they are of an operating (venting) type, commercial windows, usually relatively large, are fitted with heavy-duty hardware of simple design, built to last with minimal maintenance. For these reasons, commercial windows are rarely built of wood; steel and aluminum are typically used for their frame and sash components. When a traditional interior is desired, high-quality wood residential windows may be used.

Residential windows, while they certainly might benefit from the design features of their commercial counterparts, are generally built of lighter-duty materials, often wholly or partially of wood. They tend to be less expensive than commercial windows, are configured to install readily in a wood-frame structure, and are typically available in a range of standard sizes.

## Glass

Although a thorough discussion of the various thermal and safety properties of window glass is beyond the scope of this book, a brief review of common alternatives is presented here.

Figure 3-2 illustrates a series of glass types typically used as residential and light commercial glazing.

Common single-pane window glass has been the most widely used glazing material for centuries. Its use has declined in both residential and commercial applications, because it is neither safe nor energy efficient. Special forms of single-pane glazing—plate, float, obscure, corrugated, and mirror—are used for specialized applications, however.

**Figure 3-2**
*Glazing options.*
*(a) Single-pane glass.*
*(b) Double-glazed.*
*(c) Low-E insulating.*
*(d) Dual-space.*
*(e) Laminated glass.*

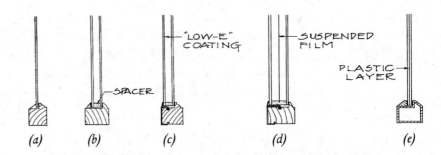

Double-pane, or insulating, glass uses a layer of trapped air as a thermal buffer. It offers good resistance to heat flow and is widely used in residential and light commercial applications. The single glass panes are joined with an airtight edge seal, which also acts as a thermal break. The composition and construction of this edge seal varies between manufacturers, and it does have an effect on the thermal performance of the overall window. Increasing the width of the space between the panes likewise increases the window's resistance to heat flow, within practical limits.

To further increase a window's thermal resistance, the space between the panes can be filled with a gas possessing increased insulating properties. Argon and krypton gasses are used by several manufacturers, although there is some debate as to the longevity of such a system. The gas, regardless of its inert state, is under a strong concentration gradient, and may diffuse away over a number of years.

Various tinting materials can be added to window glass to reduce radiant heat transfer, by simple blockage.

Silver oxide coatings, either sputtered (soft-coat) or pyrolytic (hard-coat) can impart excellent thermal enhancing properties. Termed low-emissivity, or low-E coatings, these are applied to thermal-pane glass to impart specific energy-saving qualities to the glass. They work without obstruction of view, and since they block virtually all ultraviolet radiation, interior textiles will not tend to fade. Depending on which surface of the glass they are applied to, low-E coatings can be used either to enhance heat gain and slow heat loss in cold northern climates, or to reduce heat gain in hot southern climates.

The addition of a second, or even a third, dead air space by the use of a third glass pane or one or two suspended films between the inner and outer panes, in conjunction with strategically placed low-E coatings, can reduce heat flow even more, while still maintaining a full, clear view.

For safety purposes, certain window (and door) glazings must be constructed or treated in such a way as to minimize the chance of breakage, or at least to minimize the size and hazard of broken shards. The most common safety glazing used in residential and light commercial buildings is tempered safety glass, which has been heated and rapidly cooled to substantially increase its breaking strength. It can break; however, the glass will crumble into tiny, fairly harmless cubes. Any cutting or machining operations to be performed on the glass must be completed prior to the tempering process. Tempered glazing is typically required for use in glass doors and in interior glass partition assemblies that are close to the floor. Less commonly, laminated glass is used as a high-performance safety glazing, often in overhead applications. This material, identical to automotive windshield glass, consists of a film of polycarbonate plastic fused between two panes of glass; the plastic acts as a reinforcing agent should breakage occur.

## Sash and Frame Materials

Figure 3-3 illustrates a series of material combinations that are in common use by window manufacturers.

Cheap wood windows are often single-pane with inadequate operating and locking hardware. Finger-jointed wood stock is typically used for exposed parts. These windows should be limited to economy construction, areas of infrequent use, or agricultural/industrial secondary buildings. The all-wood construction requires regular exterior maintenance. It is generally best, if specifying all-wood windows, to use reputable brands and to be sure all wood parts are treated for rot resistance prior to assembly.

Inexpensive all-metal windows, either of steel or aluminum, are also often of single-pane construction. Lacking a thermal break, they exhibit poor thermal properties. Their use should be limited to temperate climates and economy construction.

A clad-wooden window generally represents a step up in the quality of window construction. Several types of cladding are available; each type reduces the need for exterior maintenance and prolongs the likely life of the

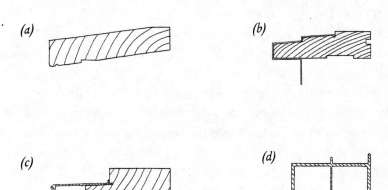

**Figure 3-3**

*Window frame materials. (a) Wood. (b) Aluminum or vinyl cladding over wood. (c) Extruded aluminum over wood. (d) All metal.*

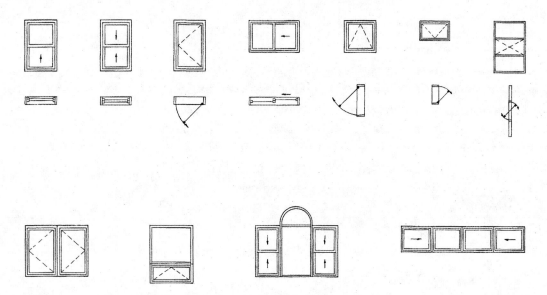

**Figure 3-4**
*Operating window sashes. (L-R, top) Single hung, double hung, casement, sliding, awning, hopper, and projected. (bottom) Combinations.*

window. Anodized aluminum is widely used for this application. In light-gauge sheet form it can be roll-formed to conform to sash or frame members; heavier extrusions can be mated with wood to form strong, durable window components. Vinyl plastic cladding requires highly specialized technology and processes. Few manufacturers have utilized this technology successfully.

## Types of Operating Windows

Figure 3-4 shows the basic types of operating window sashes. This number can be expanded greatly if one includes combinations of these basic types, such as the double casement. Each type of operating window has its place, and the decision to use a certain type must be based on factors of appearance, function, and energy efficiency.

The double-hung window (or the less common single-hung) is the window of choice in much period construction. This window type has been used in many traditional building styles for centuries. It offers excellent ventilation and good weather resistance, even when opened slightly. The sashes do not intrude into the adjacent interior or exterior space when opened. Its sliding parts are difficult to seal, though, and it generally ranks lower in terms of air infiltration resistance than some other window types. Some models are available with sashes that tilt inward for cleaning.

The sliding window is related in operation to the double-hung, but it doesn't need counterbalancing to operate properly. Its construction is simple, and it is quite effective for many purposes. Like the double-hung's, the sashes of sliding windows don't protrude when opened. These windows can't, however, be left open at all when it rains. Sliding windows don't fit many traditional styles, and they are more readily suited to commercial or contemporary designs.

The casement (and its relative, the awning) is typically rated highly in air infiltration resistance. This is because its operation and construction allow for full and complete compression of its weather stripping at the sash perimeter. Its crank mechanisms (except for those on the most inexpensive models) allow for easy opening and closing—the casement is commonly specified in residential kitchens over a countertop, where a double-hung window would be difficult to operate. Despite its excellent energy efficiency and ease of operation, the casement window has two major drawbacks. The sash, when opened, protrudes into the outdoors, and may pose a safety hazard to people outside. Also, the open sash leaves itself, as well as the entire window, exposed to the detrimental effects of weather. Occupants must remember to fully close their casement windows during storms unless they are otherwise shielded, such as by a large roof overhang. The pivot point of many modern casement windows is located several inches from the frame, allowing clearance for convenient cleaning from the inside. This may reduce the effective opening for emergency egress, however.

The awning window, like the casement, offers the potential for excellent energy efficiency and easy mechanical operation. Although it does protrude into the outdoors when opened, its configuration sheds water effectively, like a sloped roof, and allows the window to be opened during most rainstorms.

## Window Edges—Attachment and Embellishment

Once a window type, style, and brand are chosen, the specifics of installation and embellishment become very important, because poor edge detailing can render even the best of windows ineffective, inefficient, or even inoperable. As always, it is the care and forethought given to edge details that determines the success of a particular construction.

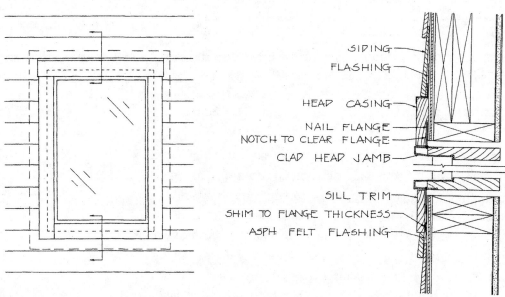

**Figure 3-5**
*A cased clad window. The exterior casing is entirely decorative, placed around the projecting window frame and fastened independently to the wall surface. The casing is rabbeted or shimmed to lay flat over the flanges.*

This section illustrates a series of exterior window details with elevation or cross-sectional views, as appropriate. Figures 3-5 through 3-9 depict a range of commonly encountered problems in window edge detailing, in both residential and light commercial work. Various methods of window installation and attachment and various trims and flashings are shown and compared.

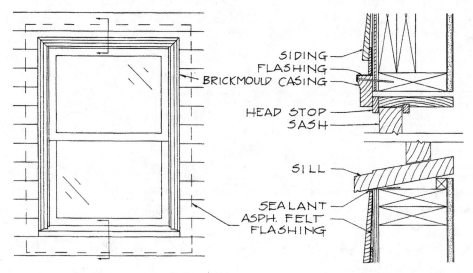

**Figure 3-6**
*A traditionally cased wood window uses brickmould or other casing to physically tie the jamb to the structure. Separate head flashing is critical.*

(a)

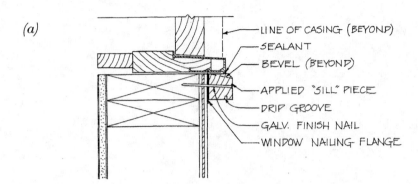

(b)

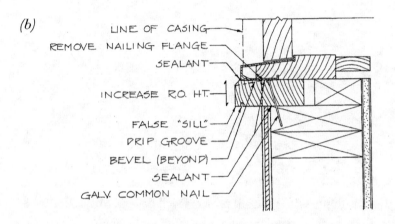

**Figure 3-7**
*Applied sill trim can be employed on a flanged window to mimic a true sill. (a) Cross-section of a sill applied to a wall. (b) A sill applied within a rough opening.*

The aluminum-clad wood casement window, shown in elevation and cross-section views in Figure 3-5, is almost always attached to the exterior wall sheathing with a metal or plastic nailing flange. This flange, which also acts as a flashing (particularly at the window head), surrounds the window perimeter; it is depicted by the small dashed line in the elevation detail. The outer dashed line is an optional, recommended flashing of asphalt felt (or similar), which lies over the flange at the top and sides and beneath the flange at the sill (see Figure 3-9). The sill flashing is then lapped onto the first full course of horizontal wood siding below. This detail shows the use of a thicker head casing, which protrudes slightly over the face and edges of the thinner side casings. This offers slightly better protection for the end grain of the side casing, and adds a classical balance to the window surround. Also shown are two methods of compensating for the flange thickness (and its effect on casing alignment): the head casing is rabbeted around the flange; the sill casing is padded with a shim. Two or three layers of inexpensive 30# felt work well here.

The more traditional window in Figure 3-6 is not clad, nor does it use a nailing flange for attachment or flashing. Its frame consists of two side jambs, a head jamb, and a true sloping sill. Jambs are typically aligned with

**Figure 3-8**
*Elastomeric sealants are commonly used as a primary seal on commercial window installations.*

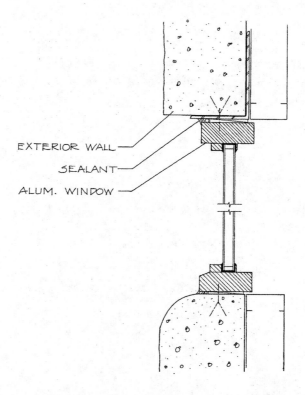

the face of the sheathing, while the sill projects well beyond for proper water diversion. The sill ends extend beyond the window opening in order to receive the lower ends of the side casings. These sill "horns" must be made of a suitable length, consistent with the casing width. This style of window is often manufactured with the casings (usually brickmould) already attached, ready for rapid installation.

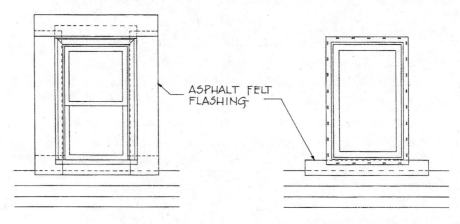

**Figure 3-9**
*The long-term exclusion of water from vertical joints at the sides of a window is best accomplished with permanent flashing. Caulking/sealing is also recommended, though it has a limited life span.*

To obtain the traditional appearance of a true windowsill with horns and casings using a modern clad or simple-framed window, an applied sill can be used as shown in Figure 3-7. Figure 3-7(a) illustrates the use of a false sill trim applied to the exterior wall beneath the mounted window. This sill-trim projects beyond the window width, far enough to receive the side casings in a manner similar to the sill in Figure 3-6. Figure 3-7(b) shows an alternate method, which uses a more substantial sill mounted within the rough opening. This method allows the sill to be completely fastened without exposed nails.

# CHAPTER 4 | Doors

A traditional entrance is a dominant exterior feature. It's often centered on the building and surrounded by elaborate moulded forms, with the entire exterior elevation and landscaping revolving symmetrically around it. Some contemporary designs downplay the concept of the entrance and do not use it as the basis or focus for the exterior design. In contemporary design, the main entrance is often concealed or camouflaged, becoming no more notable than a utility door.

Whether a grand main entrance, an abstruse alcove, or a humble back door, doors present the designer with a host of detailing problems, and excellent opportunities for creative and artistic expression. An exceptional door design may result from an adept use of edges (trims) and/or a distinctive door panel.

This chapter presents door design and detailing in the context of door field and edge conditions (see Chapter 2). The dominant field material of a door is the door panel, or slab. The door slab is commonly mounted, or hung, on a frame that allows it to move and secures the door to the wall. Door slabs may be made to function like windows, with the addition of various shapes and sizes of glass to the extent that some fullview glass doors are, in reality, windows we can walk through. Doors and their frames are made of wood, metal, plastic, or various combinations. Doors may swing on hinges or slide on tracks. (Revolving doors and overhead doors are not discussed here.)

The edges of doors—where the door meets the wall—must be detailed properly to insure good door function and lasting attractiveness. This chapter presents a range of traditional and contemporary trim constructions, as well as installation, sealing, and flashing details for typical residential and commercial doors.

# Door Construction

The construction of three typical doors is fully illustrated in Figure 4-1. Figure 4-1(a) is a diagram of a typical modern wooden residential door, mounted in a wooden frame, with an aluminum sill. Figure 4-1(b) illustrates a typical commercial door hung in a steel buck. Figure 4-1(c) is a diagram of a modern sliding glass door with a clad wood framework.

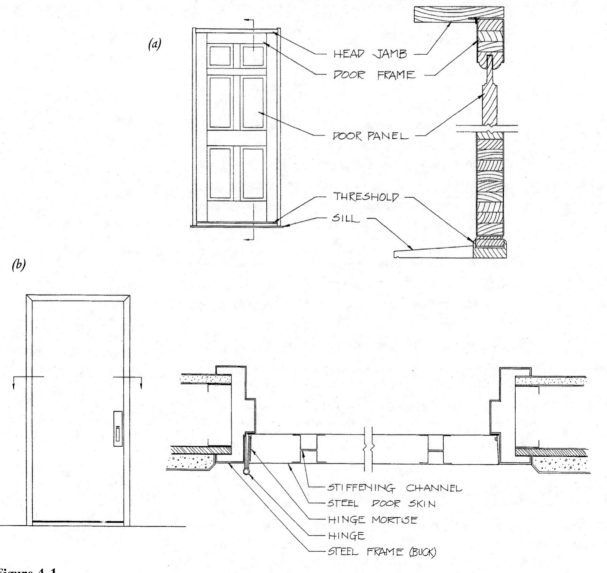

**Figure 4-1**
*The construction of exterior doors. (a) A wood residential door unit in a wood frame. (b) A steel commercial door in a steel frame. (c) A clad wood residential sliding door.*

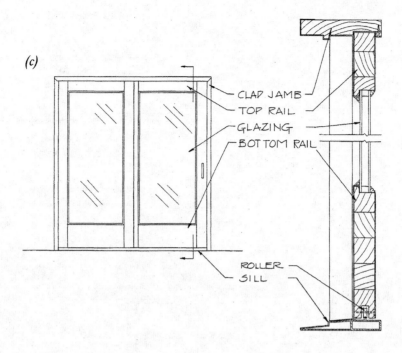

**Figure 4-1** *(continued)*

## Door Materials

### Wood

Although once the material of choice for all exterior doors, wood is used to a much lesser extent today, taking a backseat to steel. Nonetheless, wood doors of several types remain in common use, and they are certainly the choice for period reproduction work and applications where the look of wood is of primary importance.

The frame-and-panel wood door, illustrated in Figure 4-2, is a testament to the ingenuity of ancient craftsmen, who, over the course of time, developed this construction as an effective means of dealing with the problems of wood movement (see discussion below). The design persists today because it is effective and because it is what we have all come to expect a door to look like. Steel and fiberglass doors are often formed to mimic the outward appearance of frame-and-panel wood doors, even though such embellishment serves no practical purpose.

### Why Frame-and-Panel Construction Works

The use of frame-and-panel construction is prevalent throughout the woodworking industry, and with good reason: it works. It works because it is a method of construction that respects wood's natural movement in response

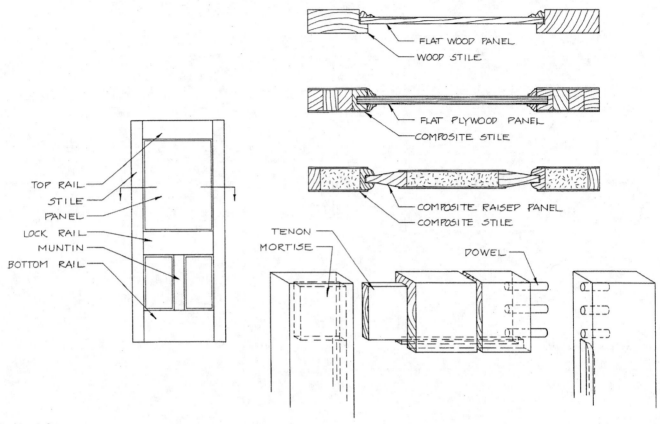

Figure 4-2
*Frame-and-panel door elements.*

to moisture content, and it minimizes the effects of this movement on the long-term accuracy of the overall construction.

The simplest way to build a door of wood is, obviously, to cut a single large board to the appropriate size and hang it in the opening. The problem with this approach, other than the difficulty in finding three-foot wide stock, is that the door will not remain the correct size, nor will it likely remain flat, due to movement across the grain and the differential movement of each face. The result is complete failure.

The next step in the right direction is the board-and-cleat construction shown in Figure 4-3(a). A series of boards is attached together with cleats placed across their faces and fastened with nails, screws, or the like. The cleats hold the boards flat and limit cross-grain movement, maintaining accuracy. The problem is that each individual board will still continue to move, and gaps will form between the boards or, worse yet, the constraint of the cleats may actually cause individual boards to split. A remedy is shown in Figure 4-3(b). The tongue-and-groove or rabbet joint allows movement

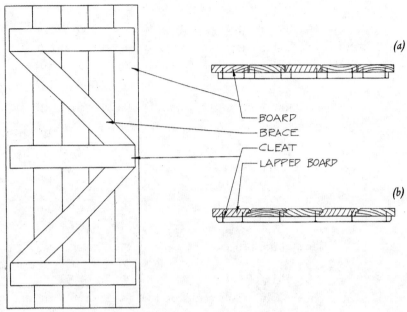

**Figure 4-3**
*A board-and-cleat door. (a) With square-edged boards.
(b) With rabbeted boards.*

without opening completely, and minimal, well-placed fasteners will prevent splitting from excess restraint. Because this door construction offers little resistance to sagging, a diagonal brace must often be fitted to keep the door square as it hangs in its opening.

The board-and-cleat door, while it can be made to work, has a few drawbacks, not the least of which is a crude, mechanical appearance. It persists today, often on barns or for certain quaint styles.

Faced with all these problems—wood movement, warping, sagging, and the need for a more finished appearance—ancient craftsmen developed the frame and panel as an effective, if complex, solution. If strong, stable particleboard, plywood, and modern adhesives had been available, they most certainly would have used them, and the familiar frame-and-panel door might never have come into common use.

Frame-and-panel construction works because the strength and accuracy of the assembly comes from a frame, shown in Figure 4-2, made of relatively narrow pieces of wood. The collective cross-grain movement of these pieces is minimal; accuracy is maintained because wood movement is negligible along the grain and therefore the frame holds its size well. The remainder of the door—the panels and intermediate frame pieces—exist mainly to fill the spaces between the main frame pieces, creating the entire door in the process.

Strength is developed at the joints between the hinge rail and the stiles. The large bearing surface of the lower rail, in particular, is totally resistant to sagging, as long as its joints are well made. Traditionally these pieces were joined using the venerable mortise-and-tenon joint; today dowels or draw bolts are sometimes substituted.

The panels are placed inside the frame with a tongue-and-groove construction or an applied moulding (see Figure 4-2). They are allowed to move freely, and have no effect on the accuracy of the frame. If the panels fit snugly, they do add to the door's rigidity; however, this is not structurally crucial if the frame is well constructed.

The design of the frame-and-panel door evolved out of need and the failure of alternative constructions. It is a triumph of material engineering, fulfilling its intended function while effectively dealing with its own problems in the process.

## *Door Choices*

The suitability of solid wood as an exterior door material is dependent on a combination of several related factors: wood species, intended finish coating, intrinsic protection from the elements, anticipated level of use, aesthetics, and renovation constraints. Unless an appropriate combination of factors favorable to wood exists, consideration should be given to alternate materials such as steel, fiberglass, or engineered wood.

The questions and answers below serve to outline the factors affecting door use and selection.

*If* . . . The use of real wood is, in and of itself, a top priority.
*Then* . . . Use real wood—choose the species and finish according to taste and budget.

*If* . . . The *look* of real wood is a priority.
*Then* . . . 1. Use real wood.
           2. Use stainable wood-grain fiberglass plastic.

*If* . . . The finish is to be opaque (painted) inside and out.
*Then* . . . Use steel, as it is virtually indistinguishable from wood under a quality paint job.

*If* . . . There is built-in protection from ultraviolet light and weather, such as a deep alcove or overhang, or there is a storm door.
*Then* . . . Use wood, if you want to—it will last longer and provide better service in a protected situation.

*If* . . . There is a need for a certain fire-resistance rating.
*Then* . . . Use steel.

*If* . . . The door will be subject to heavy, repeated, and/or careless use (apartment or commercial).

*Then* . . . Use steel.

*If* . . . Security is a priority.

*Then* . . . 1. Use steel.
2. Use fiberglass.
3. Use a thick, heavy wood such as oak.

*If* . . . The opening is nonstandard (renovation applications).

*Then* . . . Use wood, since it is easily trimmed or custom fabricated to fit.

*If* . . . Energy efficiency is a priority.

*Then* . . . Use steel with an insulating core and thermal break.

## Steel

The construction of typical residential and commercial steel door panels is illustrated in Figure 4-4.

Steel doors are available in a range of styles to match most wood door styles—raised panel, glass, and so on. These doors enjoy widespread popularity due to their low cost, long life, superior insulating value, durability, fire resistance, finish-holding ability, absolute stability and accuracy, and wide range of available styles. Steel doors solve many of the problems associated with wood doors.

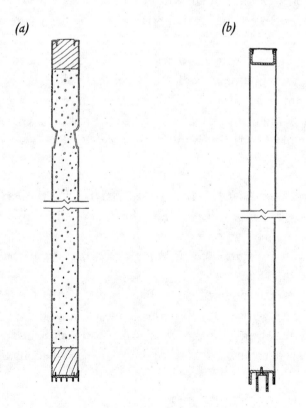

**Figure 4-4**

*Steel door panels in vertical cross-section. (a) Residential foam core, embossed, with wood edging. (b) Commercial hollow core, with steel cap and rubber sweep.*

**Figure 4-5**
*The construction of a fiberglass door panel (shown in vertical cross-section) is similar to that of a residential steel door.*

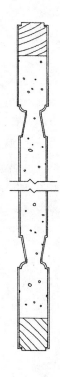

## *Fiberglass Plastic*

The construction of a typical fiberglass plastic door panel is illustrated in Figure 4-5. These doors are manufactured primarily for residential use where the appearance of wood is desired without the drawbacks of wood movement, finish failure, or low resistance to heat flow. They are moulded with a simulated wood grain—normally oak—and can take stains to varying degrees, as well as any paint finish. The finished door can very much approximate real wood, although an observer will notice the deception upon close inspection. The embossed graining is too textural to be real.

## Door Frames, Weather Stripping, and Hardware

In order for a door to perform its intended functions—opening/closing, excluding the elements, restricting air and heat flow, and providing security—it must be mated with an appropriate mounting frame. Basic frame constructions are illustrated in Figure 4-1.

A swinging door is mounted on its frame with two or more hinges. The most common hinge used today is the "butt" or "knuckle," pin-type hinge, which is about 80 percent concealed when the door is in a closed position. If the door is configured to swing outward, then the hinges must

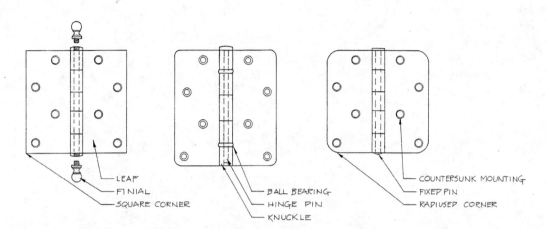

**Figure 4-6**
*Door hinges. (L-R) With optional decorative finials; with ball bearings for severe service; with fixed pin for security.*

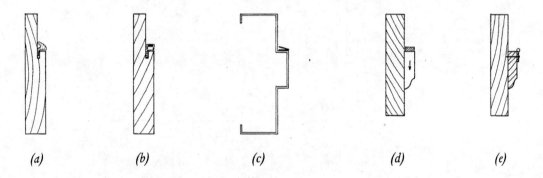

**Figure 4-7**
*Types of weather stripping. (a-e) Compressible foam; compressible magnetic; plastic leaf; adhered foam (retrofit), may require repositioning of stop; nailed bulb-type (retrofit).*

be both weather- and tamper-resistant. Several types of hinges are illustrated in Figure 4-6.

Most exterior doors are equipped with a weather strip, which, as its name implies, is a strip of material designed to seal out the weather. Many types of weather stripping are used today, ranging from compressible foam to fully magnetic. Several types of weather stripping are illustrated in Figure 4-7, shown mounted in various frames.

**Figure 4-8**
*Extended cross-section through the head of a typical residential door unit, with standard (often preapplied) brickmould casing.*

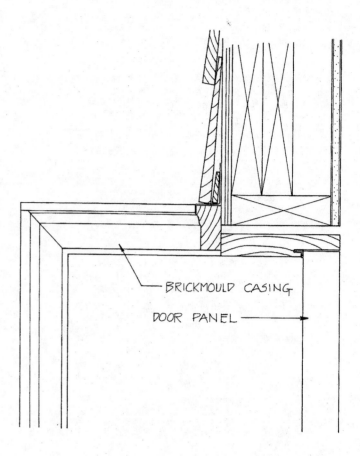

## Door Edges—Attachment and Embellishment

Once a door type, style, and brand are chosen, the specifics of installation and embellishment become very important, because poor edge detailing can render even the best of doors ineffective, inefficient, or even inoperable. As always, it is the care and forethought given to edge details that determines the success of a particular construction.

This section illustrates a series of exterior door details with elevation and cross-section views. Figures 4-8 through 4-10 depict a range of commonly encountered problems in door detailing in both residential and light commercial construction. Door installation, attachment, trim, and flashing are illustrated.

Figure 4-8 shows brake-formed metal flashing at the head casing, beneath the wood siding.

# Doors

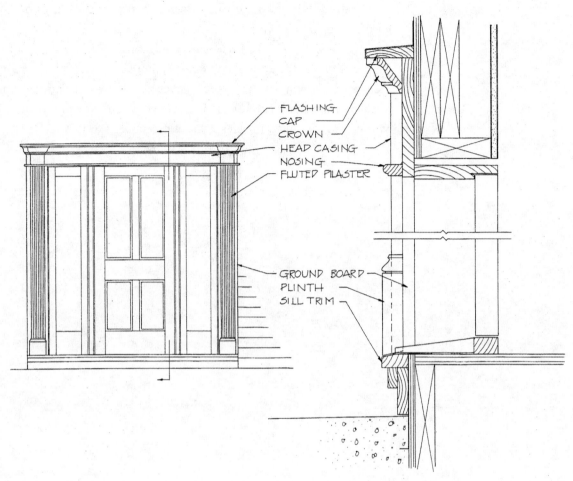

**Figure 4-9**
*A classical site-built trim scheme. The prehung door unit needs to be ordered without preapplied casing.*

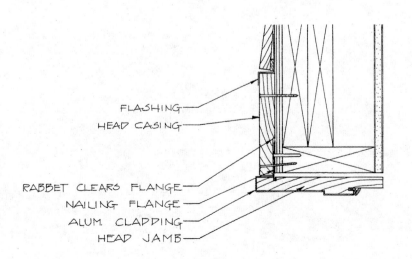

**Figure 4-10**
*A clad residential door frame, like a window frame, mounts with nailing flanges. Trim is applied around the projecting frame.*

Along with flashing, the intricate built-up mouldings of Figure 4-9 must be well sealed with caulk and paint for a lasting installation. This door assembly is shown flanked with two sidelight windows; other configurations (single door, double doors) may be so embellished with similar success. Note the use of a ground board, which serves as a flat mounting platform for the applied ornamental mouldings. The ground board is really the true, effective casing, and the fluted pilaster is a motif mounted to it.

CHAPTER 5 | # Walls and Facades

Aside from its overall form and configuration, nothing serves to define a building's exterior more than the treatment of its elevational surfaces (or *sides*). Many materials—stone, brick, glass, metal, plastic, and a variety of forms of woods, used singly or in combination—can clad the walls of a building. Each imparts a particular appearance, texture, and feeling to an exterior facade, and each, in turn, presents a host of detailing problems for the designer and builder.

From the standpoint of architectural woodwork, the most significant division of exterior cladding is *siding*—a generic term encompassing a multitude of materials, applications, and systems. Siding is any nonmasonry material used to cover and finish the exterior walls of a light frame building.

Siding may be in the form of sheet materials such as plywood, linear materials such as clapboards or vertical boards, or pieced materials such as shingles. Where siding ends, changes direction, or changes material, special trims play significant roles in creating a cohesive, integrated, weather-resistant, and attractive overall enclosure.

Since siding and exterior trim are such dominant elements on a building's exterior, significant care and forethought should be given to both the selection of materials and the details of installation, in order that the exterior system as a whole performs as well as possible. Details of edge conditions at corners, openings, and other frontiers must be designed and executed carefully.

This chapter describes and illustrates a selection of construction details for most siding systems and trim designs in common use. Traditional and contemporary applications of clapboard, vertical board, shingle, plywood, and several nonwood counterparts are covered; as well as special details such as shutters and panels. Various application alternatives are illustrated and compared, and potential trouble spots are identified.

## Siding Materials

Table 5-1 compares the characteristics of various common siding materials.

Wood sidings are generally available in redwood, cedar, and various softwoods: pine, spruce, and fir. Not all patterns are manufactured in all species, and the availability of less common styles often depends on regional building practice and market demand.

Redwood and cedar have natural extractives that make them highly decay-resistant and thus very desirable as exterior woods. Their clear grades are generally considered premier siding materials. They possess a natural beauty that is often displayed with a finish of transparent or semitransparent sealers or stains, or they can be left to weather naturally to a silver gray.

The other softwoods—pines, spruces and firs—are not naturally decay-resistant, and are often utilized when a painted (opaque) finish is desired. Their clear grades, well painted and maintained, will last many decades with little degradation. The knotty grades of any siding are more prone to finish failure (especially with surface coatings like paint) since 1) the wood in and around the knot reacts to moisture content differently than that of the surrounding clear wood, causing differential movement; 2) the knot often contains extractives that bleed through finishes, creating unsightly stains; and 3) the knot may crack or split, admitting water, which causes an increasing cycle of degradation. While a clear siding costs more initially, it will likely cost less over the life of the building due to its longevity.

Many sidings are available with both a rough (saw-textured) face and a smooth (surfaced) face, although certain patterns and grades are intended for only one face to be exposed. A textured face tends to hold its finish for a much longer time than a smooth face. Also, an even texture allows for a more even finish than a smooth face, which has uneven wood grain and porosity. A more complete discussion of the merits of textured siding is included in Chapter 17.

Extruded or formed sidings of vinyl and aluminum are among the most maintenance-free of all sidings. They are manufactured in many traditional and contemporary patterns. Vinyl sidings are usually molded in one solid color throughout, so cuts and scratches hide well. Some cheaper vinyl goods are coextruded—a layer of high-grade vinyl is laminated to a back of low-grade vinyl.

## Table 5-1 Comparison of Siding Materials

| Material | Advantages | Disadvantages |
|---|---|---|
| Redwood and cedar; clear grades | Attractive, decay-resistant, unlimited color choice, holds finish well (esp. textured face), good dimensional stability, renewable resource | Requires periodic maintenance, expensive |
| Redwood and cedar; knotty grades | Rustic appearance appeals to some, decay-resistant, less expensive, unlimited color choice, good dimensional stability, renewable resource | Rustic appearance may be objectionable, requires periodic maintenance, knots hold finish unevenly and may contribute to failure |
| Other softwood; clear grades | Attractive, less expensive (some species), unlimited color choice, renewable resource | Requires periodic maintenance, no natural decay resistance, variable dimensional stability |
| Other softwood; Knotty grades | Rustic appearance appeals to some, inexpensive, unlimited color choice, renewable resource | Requires periodic maintenance, should be painted to hide knots and to fully protect material, no natural decay resistance, variable dimensional stability |
| Vinyl | Virtually maintenance-free, long lasting, inexpensive | Highly specific application procedures must be followed to avoid failure, limited color choice, color cannot be changed, poor dimensional stability (thermal), uses nonrenewable resource |
| Aluminum | Low maintenance, long lasting | Easily dented, limited color choice, may require repainting, low dimensional stability (thermal), uses nonrenewable resource |
| Wood-fiber composite | Attractive, nearly "perfect" appearance, inexpensive, unlimited color choice (paint only), factory-primed, renewable fiber resource | Poor history of performance, highly specific application procedures must be followed to avoid failure, must be carefully monitored and maintained to prevent destructive moisture intrusion |
| Plywood | Inexpensive, excellent for smooth flat accents, unlimited color choice | Edges must be well protected from moisture intrusion to prevent glueline failure, some applications may have "budget" appearance, limited lengths may create troublesome horizontal joints |

Vinyl siding, despite its advantages, does have several drawbacks. It lacks the appearance of the traditional wood siding it mimics, either because of its sheen, embossed texture, or, depending on available lengths, highly visible lap joints. Color selection is limited, and once chosen, the color cannot be changed. It also has a high rate of thermal expansion and contraction. Correct installation, particularly proper nailing and allowance for movement, is critical. Most vinyl siding installations are not waterproof. Vertical joints and corners of openings are typically prone to leakage, and the use of a water-resistant underlayment and careful flashing is necessary. Care should be taken when using vinyl siding as a cover-up over dilapidated old wood siding. Often a serious rotting situation can be hidden, and the vinyl covering makes potentially dangerous conditions difficult to monitor.

Modern vinyl siding systems come somewhat close to having an authentic appearance, and with careful and thoughtful installation, they can look quite good at a distance. An arguable disadvantage to using vinyl siding is that its manufacture makes significant use of nonrenewable fossil resources.

Aluminum siding is formed from roll stock that has a baked-on enamel coating. Like vinyl siding, the color selection is limited, although dull-sheen aluminum sidings can be repainted as easily as wood. Aluminum is fairly inert, and aluminum siding is quite maintenance free. Unlike vinyl, however, it shows scratches, and dents easily and permanently. Aluminum is another nonrenewable resource, and it requires large amounts of energy to extract and refine. It may be most effectively used in limited amounts, as flashings and rain gutters.

Wood fiber composite sidings include the various types of hardboard and flakeboard sidings. While these are marketed under various trade names that make different performance claims, all are manufactured similarly of fine wood fibers or flakes bonded with adhesives under high pressure. Hardboard sidings possess an excellent, consistent initial appearance. They are primed at the factory and hold paint well, especially the textured styles. They make good use of mill by-products, and their manufacture utilizes highly renewable forest resources. Unfortunately, these siding materials have a poor history of performance. Improper installation and/or infrequent maintenance can cause moisture to breach the protective paint coating, causing swelling and subsequently destroying the material's integrity.

Various exterior grade plywoods and overlaid panels can be used as siding materials, in one of two ways: they can be installed as a primary siding (or siding/sheathing) material, or they can be used in a limited capacity as accent panels. Either way, the edges must be fully protected from moisture intrusion to prevent glue failure and subsequent delamination.

## Siding Styles

Figure 5-1 illustrates the range of commonly available siding styles. When the variables of species, texture, color, and pattern specifics are considered, the actual number of available sidings is far greater.

Each of the sidings in Figure 5-1 depends on some type of overlapping joint to exclude water. This joint is very effective in *horizontal* applications, since gravity drives the water down and out. The overlapping form is simple, direct, and effective, and can be found in other constructions, such as roofing. The overlapping joints of certain *vertical* sidings can lose their effectiveness if the joints become loose, since there is no guarantee that water driven down will necessarily flow out.

When one piece of siding overlaps the next, the width showing is the *exposure;* the width hidden is the *overlap*. For most siding styles, the amount of exposure and overlap is fixed—determined or gauged by the specific machined or extruded dimensions of the siding's edges. Only beveled clapboard and the (less common) board styles can be readily adjusted to fit specific conditions.

Many sidings can do double duty as sheathing, especially on low-budget and utility constructions. All sidings on wood framing, except plywood and diagonally applied wood, require structural bracing (a let-in diagonal wood brace or steel tension-compression bracing) in the absence of structural sheathing beneath the siding.

Of all the siding styles illustrated in Figure 5-1, clapboard, or beveled, siding (and its imitators of vinyl, aluminum, or wood fiber composites) is the most common, recognized, and widely-used pattern of all horizontal sidings—or of all sidings, for that matter. Clapboard is what comes to mind when siding is mentioned.

Wood clapboard is available in several widths and thicknesses, as well as a variety of grades. The thinner patterns ($1/2 \times 6$ or $1/2 \times 8$ in., measured at the lower edge) can abut thinner trims (corners, casings), since their maximum installed thickness is generally less than an inch. Heavier patterns ($3/4 \times 8$, $1 \times 10$ in.), however, must use thicker or built-up trims since their installed thickness can exceed one inch.

The interplay between a material's thickness and its amount of exposure has strong design ramifications; widely varied appearances are possible due to variations in the size and frequency of the siding's *shadow line*. The shadow cast by one piece of siding onto its neighbor below has a visual impact; the strength and prominence of this impact is directly dependent upon material thickness and

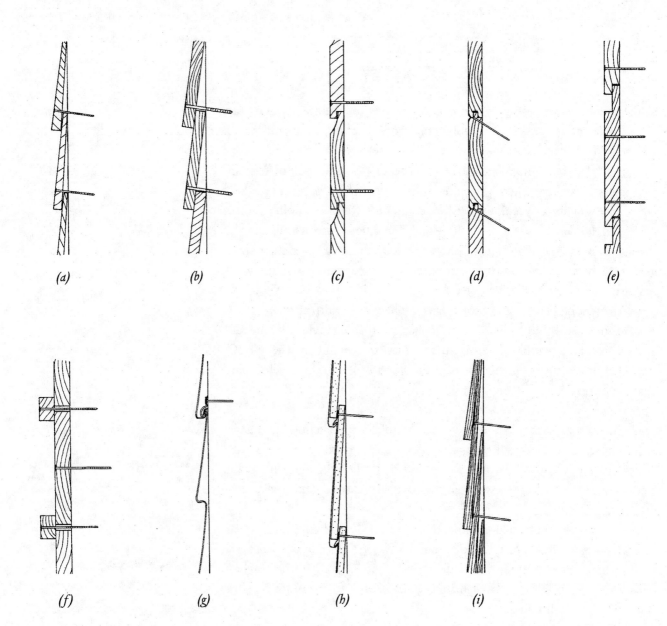

**Figure 5-1**
*Siding patterns. (a) Bevel clapboard. (b) Board. (c) Rabbeted lap. (d) Tongue-and-groove. (e) Channel. (f) Board-and-batten. (g) Vinyl clapboard. (h) Aluminum clapboard (with backer). (i) Wood shingle. a–c and g–i are installed horizontally; d and e can be installed horizontally or vertically; f is suitable for vertical installation only.*

# Walls and Facades

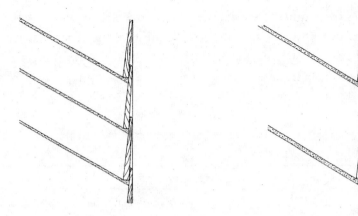

**Figure 5-2**
*Siding thickness and exposure affects shadow line size and frequency.*

exposure. The refined narrow siding of a traditional New England home imparts an entirely different feeling than that of a ski chalet clad in overlapping 1 × 10 inch boards. Generally, the narrower and thinner patterns are associated with more traditional architecture, while the wider and thicker patterns are associated with more contemporary or rustic styles (see Figure 5-2).

Clapboard and other board sidings are unique in their ability to be adjusted to fit the circumstances of the installation. For example, while the recommended exposure for ½ × 6 inch bevel siding is 4½ inches, a skilled installer may reduce or increase the exposure if it allows for a neater installation. Figure 5-3 illustrates the advantages of adjustable exposure in creating a neat finished installation.

The remaining siding styles shown in Figure 5-1 are self-gauging; that is, each course fully depends on the previous one and its respective millings to determine overlap and exposure. Channel siding, which has a slightly larger

**Figure 5-3**
*The adjustability exposure of clapboard allows for better alignment with window trim.*

overlap than some other wood patterns, and vinyl siding, which is flexible and can be "stretched" up and down a bit, are slightly adjustable, but not nearly to the extent of clapboard.

Table 5-2 summarizes the advantages and disadvantages of each primary siding style.

## Table 5-2 Comparison of Siding Patterns

| Pattern | Advantages | Disadvantages |
| --- | --- | --- |
| Bevel clapboard | Fits wide variety of styles, adjustable exposure for best fit, sheds water well | Requires considerable time and skill to apply correctly, follows wall imperfections |
| Board | Rugged standout appearance, adjustable exposure for best fit, rides over wall imperfections, sheds water well | Requires thick stops and trims, difficult to restrain |
| Rabbeted lap | Attractive and unique styles, rapid gauged installation, sheds water reasonably well | No adjustability, difficult to match as patterns change over time |
| Tongue-and-groove (horizontal) | Allows blind nailing, rapid gauged installation, sheds water reasonably well, can double as sheathing | No adjustability, difficult to match as patterns change over time |
| Channel (horizontal) | Strong deliberate linear appearance, rapid gauged installation, sheds water reasonably well, may double as sheathing | Limited adjustability, difficult to match as patterns change over time |
| Tongue-and-groove (vertical) | Allows blind nailing, rapid gauged installation, covers curved surfaces well, can double as sheathing | No adjustability, requires horizontal blocking, may not keep all water out |
| Channel (vertical) | Strong deliberate linear appearance, rapid gauged installation, may double as sheathing | Limited adjustability, requires horizontal blocking, may not keep all water out |
| Board-and-batten | Rustic appearance appeals to some, adjustable exposure for best fit, can double as sheathing | Rustic appearance may be objectionable, requires horizontal blocking, may not keep all water out |
| Vinyl (extruded) | Rapid gauged installation, covers curves well, floats over wall imperfections, sheds water well, prefinished | Slight adjustability, unattractive lap joints are highly visible, patterns and colors change over time |
| Aluminum (roll-formed) | Rapid gauged installation, sheds water well, prefinished | No adjustability, unattractive lap joints are somewhat visible |
| Wood shingle | Attractive textural appearance, highly adjustable exposure, sheds water well | Requires considerable time and skill to apply well |

**Figure 5-4**
*Wood shutter installation.*

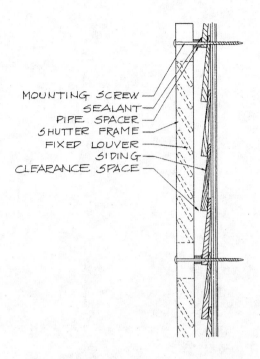

## Shutters

Hinged wooden shutters, originally used as protective coverings over windows to shut out weather or intruders, are often specified on certain traditional facades for historical accuracy or to improve visual interest or balance. Modern shutters are seldom installed to operate; rather they are fixed in place at either side of a window in a permanently open position.

Slatted or paneled wood shutters should be used over wood sidings; they should be fastened with rustproof screws over spacers, as shown in Figure 5-4. The space allows for free drainage and air circulation. Slatted shutters that are permanently attached in place do not require movable slats, though, in the interest of authenticity, the slats should be oriented in the upward position (when viewed from the front) in deference to their supposed downward position should the shutters be closed. Figure 5-4 illustrates the correct orientation. Vinyl plastic shutters, appropriate only for use over vinyl or aluminum siding, should be fastened with long screws to the underlying substrate through oversized holes in the siding (to allow movement).

The finest wood shutters are constructed of cedar, cypress, or redwood, with good joinery, for long service. All wood shutters, since they are generally of light construction with many parts, require frequent maintenance to ensure long service.

**Figure 5-5**
*The parts of a siding installation.*

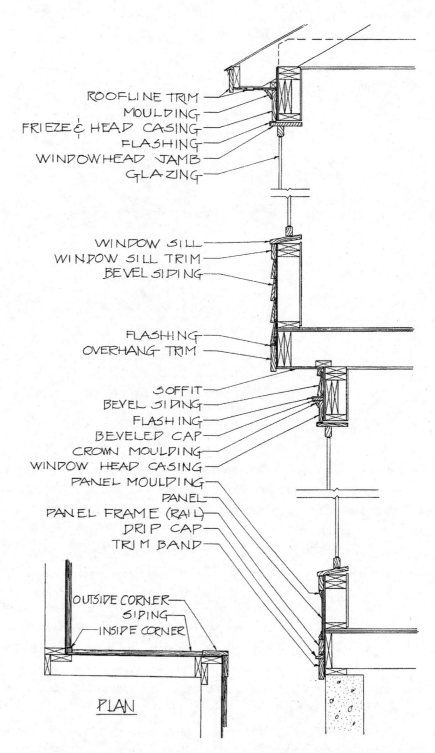

Shutter size should be appropriate to the window in question: the height should correspond to sash height (not the overall window height), and the width should be one-half the window width. Shutters should be sized and mounted so as to appear that they could close and effectively cover the window.

# Walls and Facades

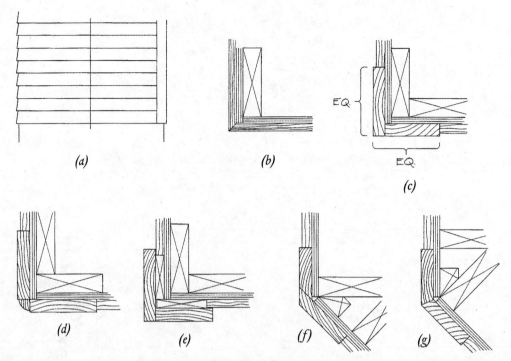

**Figure 5-6**
*Outside corner details. (a [right], b)Mitered corners minimize vertical joints, yet lack definition and are more difficult to execute. (c) Corner boards are typically sized for symmetry. (d) The use of a filler moulding. (e) A built-out corner overlaps clapboard siding, but it creates many small spaces that attract insects. A lapped joint, which is more forgiving of wood movement than a mitered joint (g).*

Figure 5-5 illustrates all the parts in an exterior wall siding system. Most designs use only some of these many options. Additional construction details for residential and light commercial exterior walls are illustrated in Figures 5-6 through 5-8.

Figure 5-5 shows a cross-section of a variety of exterior wall finish details, showing transitions and ties to the foundation, windows, overhang, and roofline.

Figure 5-6 illustrates several outside corner options. Details (c) and (f) are recommended for their simplicity and durability. Water exclusion at vertical joints, like those at corners and windows, is best achieved with a combination of two methods: flashing and sealing. A strip of saturated felt or a water-resistant building wrap will keep seepage from wetting the underlying

# CHAPTER 5

**Figure 5-7**
*Elevation and cross-sections of a paneled window wall. Flat or raised panels can carry a repeating window motif onto the wall field.*

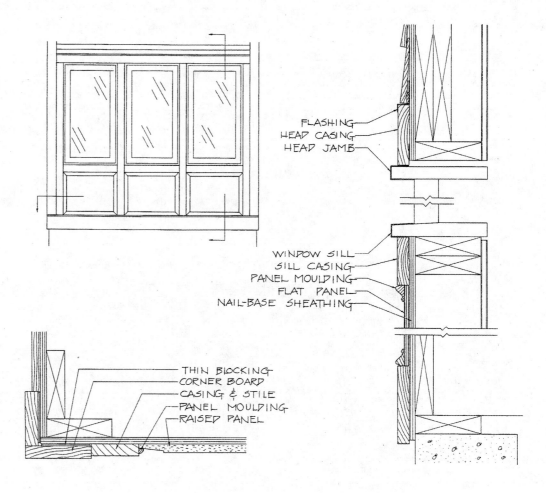

**Figure 5-8**
*Shutter alignment. The right shutter is correctly installed.*

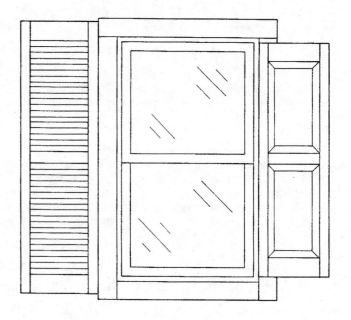

structure. This flashing is not perfect, though, since the siding and trim nails penetrate it. A thin bead of high-grade sealant or caulk should be applied to the vertical joint as further insurance. Remember, however, that the success of an applied protection such as caulking depends on proper periodic inspection and maintenance.

Paneled exterior walls (Figure 5-7) should be designed to exclude water. Since there are few naturally overlapping elements in a panel assembly such as this, the use of flashings at transitions is important, as well as the use of embedded and/or applied sealants at all other potentially loose joints. Note the use of a thin blocking to create a step, or "reveal," from the corner board to the casing. The use of thicker stock would have the same effect.

CHAPTER 6 | # Roofline Trim

Woodwork that is installed between the termination of the roof and the upper edge of the exterior wall of a building—the edge between the roof and wall fields—can be termed roofline trim. It can assume a variety of forms, from a simple linear accent to an intricate assembly of classical moulded forms. Nearly all buildings have roofline trim to some extent, and while it is certainly possible to construct a building without any definition at its roof's extremity, the results may be unsatisfactory and are usually impractical.

The aesthetic and functional success of a roof design is influenced by the delineation and performance of the roofline trim. A small change in dimension or aspect can have a noticeable impact on the building's finished appearance. The overall design category of certain buildings is defined, to a certain extent, by the scale and complexity of this work. The roofline trim is the visual "tie" between roof and wall; it influences our perceptions of roof mass, wall height, and building presence.

Regardless of its level of complexity, the roofline trim must be carefully detailed to function properly. In the same manner that the ancient elemental cornice preserved the fragile clay walls below, roofline trim can protect (and sometimes ventilate) a modern wood building. In addition, it should be constructed to hold up under demanding conditions. Roofline trim is generally exposed to large amounts of sunlight, rain, and wind; it may also suffer occasional battering from carelessly positioned ladders and deterioration from pest activity. It is often difficult to reach and tedious to maintain, so it should be designed to perform well with infrequent attention.

# Trim Components

Figure 6-1 illustrates cross-sections of typical roofline trim at two fundamentally disparate locations on a sloped-roof building: the eave and the gable. The entire trim assembly at the eave is equivalent to a classical entablature (see Figure 2-2), although few modern buildings contain all the elements of a complete entablature. The overhanging portion of the roofline trim is the cornice, usually composed of a fascia (from the Latin for horizontal band), often a soffit, and sometimes additional moulded accessories. The wall immediately beneath the overhang can be trimmed with a frieze element and, immediately below that, an architrave. The architrave is really a derivative of a classical beam; it is a common functional roofline trim element on open post-and-beam structures like porches.

Much modern roofline trim is, of course, highly simplified to reduce its cost and maintenance; it is often constructed of only a simple fascia and soffit. The fascia is the element that is most visible in elevation. The plane of the roof rests, in effect, on the fascia's uppermost edge. Some differentiate between eave and gable fascias by assigning the term "rake board" to the sloping fascia at the gable. I will use the terms "eave fascia" and "rake fascia." The two are generally of the same material on any given building, although they may vary in width as a result of roof pitch geometry or to achieve a certain

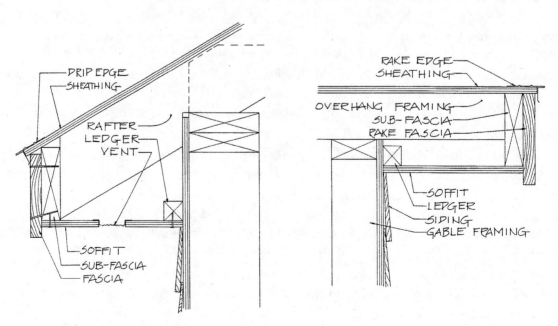

**Figure 6-1**
*Cross-sections of typical roofline trim at the eave (left) and gable.*

design effect. The face of the rake fascia is nearly always installed plumb, while the eave fascia may be installed plumb, perpendicular to the roof pitch, or, rarely, at some other critical angle.

If the roof overhangs a sidewall, its underside is the soffit. The soffit may be left open, or it may be closed in a number of ways. Roof ventilation ports are often incorporated into the design of the soffit. In a manner collateral to the fascia, an eave soffit may be level or parallel to the roof pitch. A rake soffit is nearly always installed parallel to the plane of the roof.

## Design Considerations

A good roofline trim system should satisfy several requirements:

**1.** It must provide a measure of protection for the wall below by blocking ultraviolet radiation and by carrying water away from the wall where it can drip free of the building.

**2.** It should, in most wood-frame construction, provide for adequate ventilation of air to the underside of the roof sheathing.

**3.** It should conceal and protect from deterioration vulnerable framing members of the roof structure.

**4.** It should, except in the case of suspended or concealed gutters, provide a base for the secure attachment or bearing of the roof drainage system.

**5.** It must satisfy the aesthetic needs of the building's overall design

Since the first four, practical considerations can be readily met with a variety of materials and assemblies, the fifth concern should take priority in the design of a roofline trim system. First develop a visual concept, and then verify that the functional needs are met.

The scale and complexity of the roofline trim can have a sizable impact on a building's overall appearance. For example, consider the common contemporary exterior shown in Figure 6-2. The sloping soffit and wide overhangs give the desired appearance of a slab laid atop the walls; unfortunately, the four-inch-wide fascia makes the roof appear thin and flimsy. Figure 6-3 shows the same building with a nine-inch fascia: the roof assumes a commanding and dramatic presence.

The gabled exterior in Figure 6-4 is a good example of traditional New England roofline trim—sparse and austere. Extending the eave and gable overhangs and terminating the corners with hipped returns (Figure 6-5) gives the building a distinct elegance and provides a measure of protection for the sidewalls.

**Figure 6-2**
*A narrow roofline trim can appear weak.*

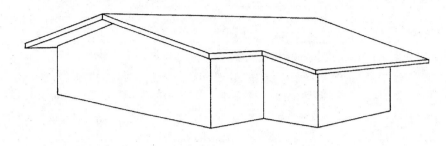

**Figure 6-3**
*The addition of a wide fascia adds substance.*

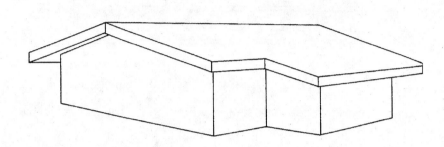

## Materials

The choice of materials for roofline trim depends on two factors: 1) the materials used to clad and finish the exterior walls and 2) maintenance considerations. Typically, buildings with wood siding will have wood roofline trim, and those with metal or vinyl siding will use similarly low-maintenance materials as roofline trim. The materials can be mixed, however, to achieve desired design intents with good results. The construction details at the end of this chapter illustrate hybrid installations.

The choice of a roofline trim material depends on considerations similar to those involved in selecting siding, as outlined in Chapter 5. The range of

**Figure 6-4**
*A minimalistic roofline trim.*

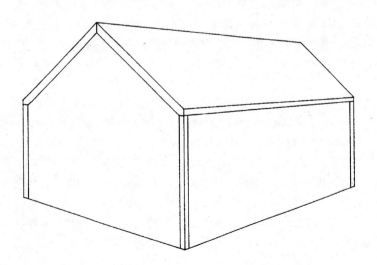

# Roofline Trim

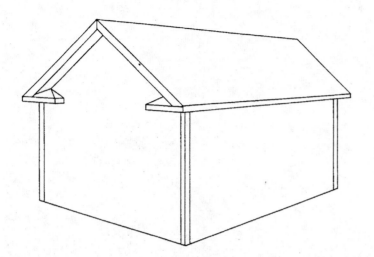

**Figure 6-5**
*Roofline complexity can add grace and refinement.*

available materials for roofline trim is similar to that for wall siding, and priority should be given to materials that provide good service and appearance.

## Roofline Trim Details

Figures 6-6 through 6-10 illustrate a series of roofline trim construction details. Figure 6-6 illustrates two versions of no-maintenance roofline trim assemblies. The one shown in Figure 6-6(a) is common in new residential construction;

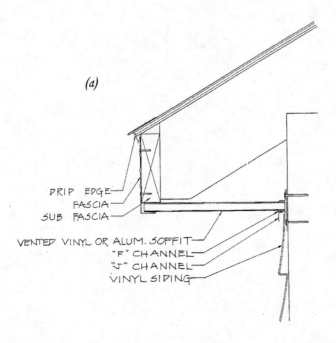

**Figure 6-6**
*Maintenance-free roofline trim constructions.*
*(a) New construction.*
*(b) Retrofitting an older residence. Light-gauge aluminum trim (fascia, here) is best installed during warmer weather to minimize expansion deformation.*

**Figure 6-6**
*(continued)*

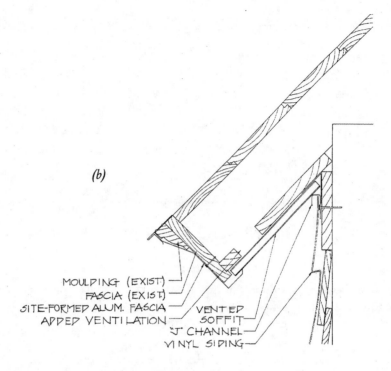

**Figure 6-7**
*Open soffits. (a) The use of tongue-and-groove decorative sheathing may require a transition wedge to blend with thinner sheathing plywood up the slope. (b) Finish-grade exterior plywood provides good appearance for exposed portions.
(c) Rafter tail decoration for fully open soffits. Note that a small fascia element is retained to stiffen the roof edge.*

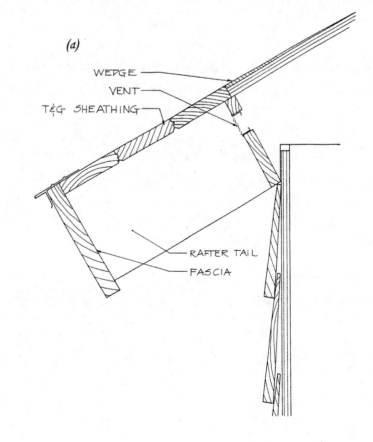

# Roofline Trim

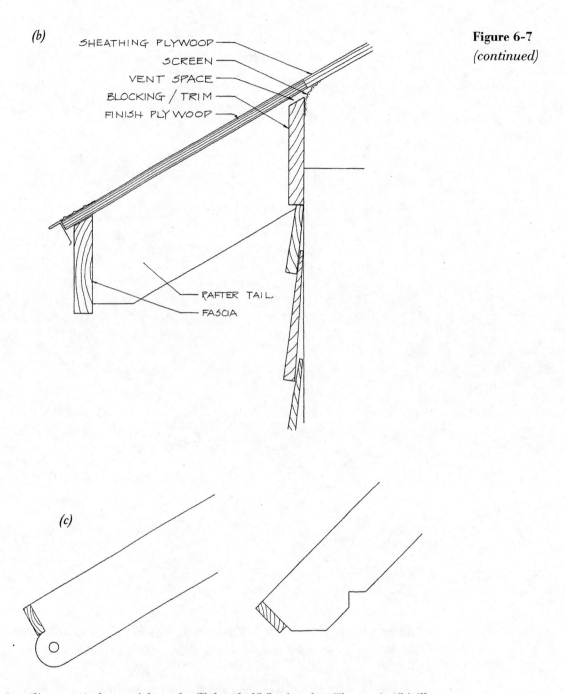

**Figure 6-7**
*(continued)*

it utilizes typical materials and off-the-shelf fascia trim. Figure 6-6(b) illustrates a retrofitted trim system commonly employed on older residences in residing renovations. The fascia is brake-formed on-site to closely conform to the existing trim profile. The use of site-formed trim cladding instead of stock shapes can add distinction to both old and new work by offering a custom-made appearance and fit. The existing soffit board can be drilled or cut to allow ventilation. Since perforated soffit panels are not as attractive as

**Figure 6-8**
*A hybrid construction using a maintenance-free soffit with a wood fascia maintains a consistent appearance with wood siding in elevation view.*

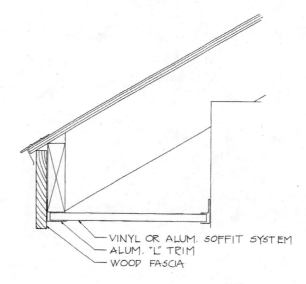

**Figure 6-9**
*A close-cornice detail using a specially designed drip-edge flashing to provide an inlet for ventilation.*

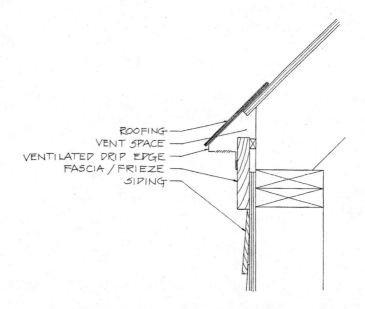

solid ones, highly visible gable soffits (not shown) do not necessarily need to be of the ventilated type, so long as the eave ventilation extends out to the gable soffit area.

Figure 6-7 illustrates two approaches to the construction of open soffits. Open soffits may be employed in order to match an existing design or simply as a way of adding visual interest to the roofline area. Figure 6-7(a) illustrates the use of a fascia set perpendicular to the roof plane, with appearance-grade tongue-and-groove sheathing visible from below. A wedge piece provides a smooth transition to the main plywood sheathing, which is often thinner. Figure 6-7(b) illustrates the use of a plumb fascia and appearance-grade plywood where it is visible. No wedge is necessary in this case. Both

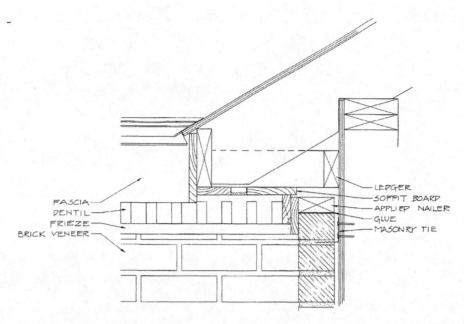

**Figure 6-10**
*Extended section through roofline trim on brick-veneer wall.*

designs illustrate the use of ventilation in the blocking area. The blocking is necessary to close off the attic area and as an insulation stop. These nonstandard constructions must be planned and executed carefully in order to look good and perform well. Figure 6-7(c) shows custom-cut rafter tails, which can be employed in certain period-style open soffits.

Figure 6-8 illustrates a hybrid roofline trim system. A wood fascia is used to coordinate with the wood siding, but a vinyl or aluminum soffit is used to provide ventilation and for cost-effectiveness and low maintenance needs.

Figure 6-9 illustrates a solution for ventilating a close cornice (no overhang). The manufactured drip edge provides built-in ventilation.

Figure 6-10 illustrates a traditional wood roofline trim assembly on a brick-veneered wall. Blocking is provided for the attachment of frieze elements.

CHAPTER 7

# Porches and Decks

Outdoor living spaces go by many names: portico, balcony, porch, veranda, terrace, patio, deck, court, piazza. If we narrow this field to structures made primarily of wood we are left with the porch, deck, veranda, and balcony. Considering that a balcony is simply an elevated porch or deck and a veranda is essentially a regional variant of the porch, our discussion comes down to two basic constructions—the *porch*, which has a roof, and the *deck*, which does not. Each of these structures presents the designer with its own set of special detailing problems—the most significant being the protection or preservation of woodwork that is subjected to the weather.

## Porches

A porch, which comes from the Latin words *porticus* (entrance) and *portus* (haven) is, in simplest terms, a covered entrance to a building, projecting from a wall and usually having a separate floor and roof. Throughout history porches have appeared in many forms, in some cases evolving to become significant architectural features dominating the exterior design of a building. A porch may be a tiny covered stoop at the back door or an expansive surround replete with multiple stairways, an elaborate balustrade, and a classical colonnade of detailed posts and auxiliary woodwork.

Figure 7-1 illustrates the essential elements of traditional porch construction.

# CHAPTER 7

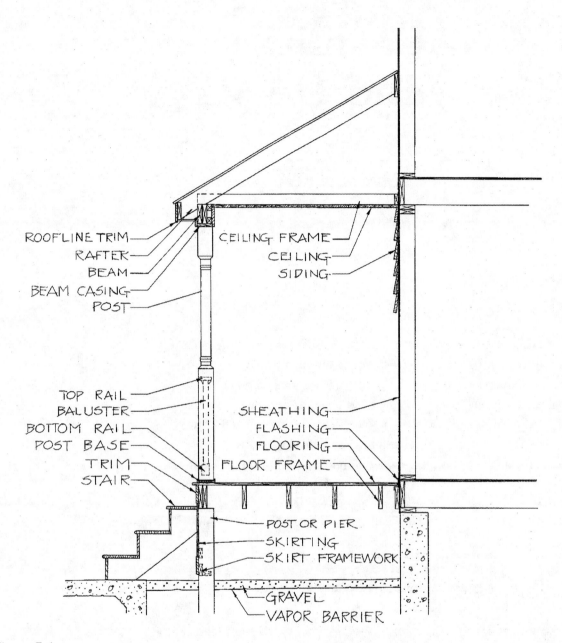

**Figure 7-1**
*The elements of traditional porch construction.*

Whether large or small, sophisticated or simplistic, the design considerations for wood porches are similar:

**1.** The roof is normally supported by the floor structure through a system of posts and beams.

**2.** The roof provides exceptional protection from the elements (particularly ultraviolet radiation), which allows for a high degree of fine detail in the woodwork of the porch.

**3.** Some precipitation will inevitably invade the porch space, and allowances must be made for its rapid removal.

**4.** The best porch designs are those that visually tie the porch to the main structure, using repeating motifs and a smooth transition of form. A porch should not appear "tacked on" as an afterthought.

## Floor and Foundation Details

From a structural standpoint, the porch is a fairly simple construction. It gains roughly half its support from the structure it is attached to, and the other half from an outlying foundation system at its perimeter. This foundation may consist of piers or posts, or possibly a continuous masonry stem wall if the floor is to be concrete (see Figure 7-1). Care should be taken to ensure that the outlying foundation is as stable and reliable as that of the main structure, lest the porch settle differentially and acquire that all-too-familiar sag of so many poorly supported porches of old.

Regardless of the material or means of support, it is imperative that the porch floor be designed to slope away from the main building in order to guide water away and off the porch. Other horizontal elements should be level (railings, ceilings), but the floor should be built with a slope of $1/8$ to $1/4$ inch per foot.

If the floor is to be of wood, the framing members should be placed parallel to the main building, enabling the porch flooring to run perpendicularly, along the direction of the slope. In most cases this can be easily accomplished with no additional framing expense, and this construction creates two distinct advantages. First, if the flooring develops any cupping, ridges, or gaps, these defects will not tend to impede the free flow of water, because they will occur along, rather than across, the path of flow. Second, except in the case of "side loading" porches, the flooring will be aligned with the flow of foot traffic from stair to door, allowing for an extremely durable end-grain "nosing" at the porch's edge (see Figure 7-2) and

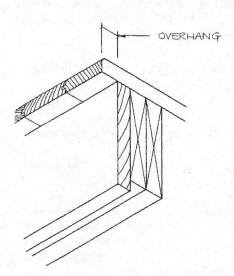

**Figure 7-2**
*The end of flooring can form a durable nosing.*

minimizing the chances of tripping on the wood defects (cupping, ridging) mentioned previously. Of course, this ideal arrangement isn't always possible or practical, but when the choice presents itself, the designer should be aware of the functional consequences of flooring layout.

Except for the case of a concrete or masonry floor system, a porch floor should be constructed of tongue-and-groove wood flooring. The interlocking tongue-and-groove assembly creates a smooth, continuous, unwavering surface that befits the formality of a porch. Gapped boards, like those on a deck, are somewhat inappropriate (and unnecessary) for a covered structure, especially at a main entrance. The exception would be the case of a partially covered deck, where the aim is to keep the flooring consistent across the differing spaces.

Several species and types of porch flooring are readily available and are summarized in Table 7-1.

Ideally, porch flooring should be blind-nailed with deformed-shank galvanized flooring nails, and sanded flat prior to finishing, in the same manner as an interior floor. This initial sanding levels any irregularities in the floor, and enables the best possible finish right from the start. The boards will, over the life of the porch, tend to move (cup, separate, warp) to some extent, depending on the amount of direct and indirect moisture-related

## Table 7-1 Porch Flooring Materials

| Material | Advantages | Disadvantages |
| --- | --- | --- |
| Redwood (quartersawn) | Excellent dimensional stability, decay-resistant, takes finishes well | Expensive, somewhat prone to denting |
| Douglas fir (quartersawn) | Excellent dimensional stability, wears well | Somewhat expensive, no natural decay resistance, grain may show through paint |
| Redwood (plain-sawn) | Decay resistant, fair dimensional stability, takes finishes well | Prone to denting |
| Douglas fir (plain-sawn) | Fair dimensional stability | Wears unevenly, grain may lift, no natural decay resistance |
| Yellow pine (plain-sawn) | Hard, wears well | Poor dimensional stability, no natural decay resistance |
| Yellow pine (CCA treated) | Hard, wears well, decay-resistant | Poor dimensional stability, requires kiln drying after treatment for decent installation, treatment may affect finish material choice and timing of application |

# Porches and Decks

stress they are subjected to.

Wood movement can be kept to a minimum by following several precautions. These recommendations apply to all exterior wood flooring and should be followed for a first-class installation:

**1.** Always acclimate the wood to its intended installed location by allowing it to sit on-site—covered, well off the ground, and with free air circulation—for at least a week prior to installation.

**2.** Prefinish the back and edges of each board prior to installation. Once the final, top finish is applied, each board will then be completely finished and will lose and gain moisture evenly from all sides, reducing the likelihood of cupping and warping.

**3.** Install an effective vapor barrier on the ground beneath the porch, and be sure to allow for adequate drainage and ventilation beneath the porch so that the underside of the porch remains dry.

**4.** Finish (seal) the top of the porch floor and all end-grain surfaces as soon as possible after installation. If the floor will be subjected to construction traffic, then forgo sanding and paint on an inexpensive oil primer as a temporary seal. Once construction is complete, sand the floor bare and smooth, and apply the final finish.

**5.** Use a flooring type with an inherent resistance to wood movement–related defects. Quartersawn fir and redwood are ideal, although your budget may preclude their use. Plain-sawn redwood or fir are good second choices. A narrow pattern—2 to 3 inches wide—will keep the actual movement per board to a minimum and will reduce the size of potential gaps between boards. However tempting, wider boards should not be used as porch flooring. Excessive width requires face nailing (a defect), and the total seasonal movement will be unsatisfactory for the formal woodwork of a porch.

## Post Details

A very common, highly visible malady suffered by many open porches on older wood-frame homes is structural degradation resulting from the rotting of post bases. The raw, open end-grain of wooden posts, sitting directly on the porch flooring material, literally wicks water up into the posts, where it creates conditions favorable for wood rot. Damp wood supports the growth of decay-causing microorganisms and breaks paint bonds, causing finish failure, which aggravates the condition.

Figure 7-3 illustrates a common method of post reconstruction. The rotted lower section is removed and replaced with new stock joined with a strong doweled lap joint. Methods such as this have varied success, and a renovation budget might be better spent on complete post replacement. The

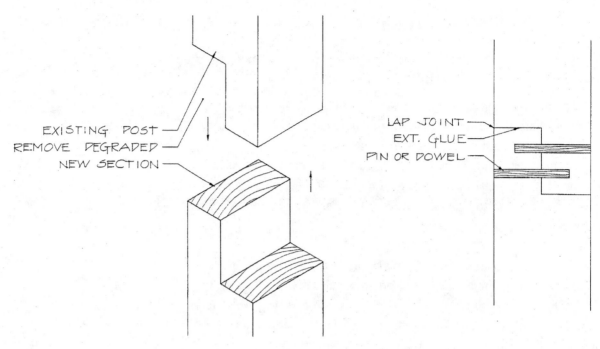

**Figure 7-3**
*A remedial repair to a degraded post.*

replacement post must, however, be detailed properly to avoid a repeat of the original failure.

Whether for new construction or as part of a porch renovation, the detailing of the porch support system is crucial to an attractive, lasting construction. Since several good options are available to assure a successful installation, the first priority should be the aesthetic considerations involved in the overall porch (and building) design. Choose a post or column pattern for its design appropriateness. A Victorian scheme often calls for lathe-turned posts with squared ends to which a balustrade and fretwork is mounted. A classical design often calls for columns patterned after a classical order, standing alone with little other attached work. Avoid the temptation to substitute an inappropriate component for its decay resistance alone, such as a crude pressure-treated timber or an off-the-shelf treated turned post, if the overall design is not well served.

Figure 7-4 illustrates several details that address the problem of degrading post ends. Note that common pressure-treated components are not a viable alternative, for two reasons:

**1.** There is a strong temptation to simply rest the post on the porch floor, since rot is ostensibly no longer a concern. Water will, however, seep up the post as readily as ever, where it can break the finish bond and cause paint peeling or flaking.

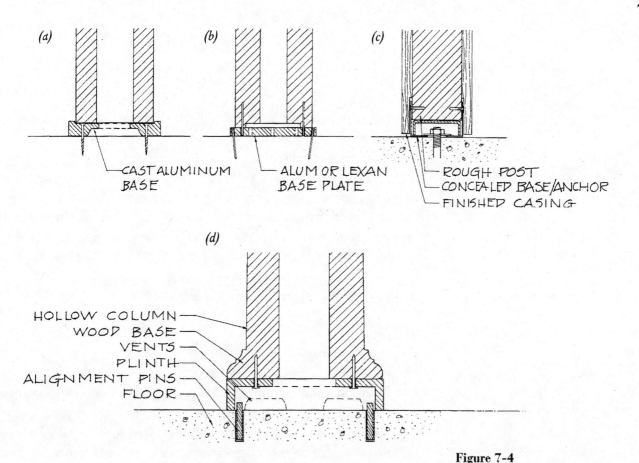

**Figure 7-4**
*Post bases. (a) A manufactured cast base, sized for standard posts. (b) A fabricated base with holes and slots for free air circulation. (c) A concealed metal base/anchor for cased posts—excellent when uplift resistance is required. (d) A ventilated cast base for large columns, aligned with pins and secured by gravity.*

2. Many pressure-treated porch components, while they certainly won't rot, are of inferior materials. Unstable, often knotty and improperly dried southern yellow or red pine in large, solid sections is a poor choice for fine exterior woodwork.

Chemical impregnation alone is not a substitute for good design. The proper use of appropriate construction details allows for broad design versatility as much today as ever, and modern porch designs should not be limited because of the availability of a small selection of pressure-treated components.

Except in the case of aluminum or fiberglass columns, the following recommendations are important to assure a lasting porch construction:

1. Create an air gap between the post end and the porch floor. Use a metal base, which will provide both adequate load transfer and good ventilation. If wetted, the post must be able to dry freely. Porous bases, of wood or masonry, delay the drying process, and are not recommended.

2. Use a post material that is appropriate for the design constraints. Use clear redwood or cedar if a clear or stain finish is intended. Use a fine-grained wood with minimal defects, well seasoned and finely milled, if a painted finish is desired. White pine or fir are good choices. Posts made of glued sections (laminated or staved, not finger-jointed) are recommended

where the posts are to be of any appreciable thickness, as they are inherently more stable and less likely to split or check. Be sure the glue used is rated for exterior exposure.

**3.** As extra insurance against failure, treat the bottom ends of wood posts with a soaking application of a wood preservative and water repellent. Follow this treatment with a brush application of the same solution over the entire post surface, so that the final coats of finish take evenly. Decay-resistant species to be stained or sealed need only a soaking application of the intended finish material. In any case, the goal is to close the pores of the end grain as effectively as possible.

## Porch Construction Details

Figure 7-5 illustrates a series of details that provide solutions to the problem of closing and finishing the roof and ceiling structure of several porch framing scenarios. The interplay of post, beam, rafter, and joist creates the groundwork, against which a multitude of trim styles may be applied. Furring, blocking, or other false framing is often necessary to achieve desired proportions, intersections, and dimensions. As with any exterior woodwork, provisions must be made to exclude and divert water. Flashings, sloped members, and proper sealing all play important roles in assuring the longevity of the structure.

Figure 7-6 illustrates basic construction details for porch balustrades, stairs, skirtings, and other elements.

**Figure 7-5**

*Porch roof and ceiling details. (a) Plywood soffit and ceiling, with protruding beam. (b) Vinyl panel soffit and ceiling, with semiconcealed beam. (c) Tongue-and-groove wood ceiling/soffit with fully concealed beam*

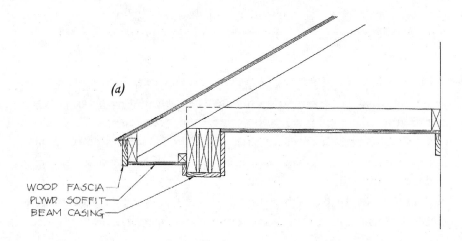

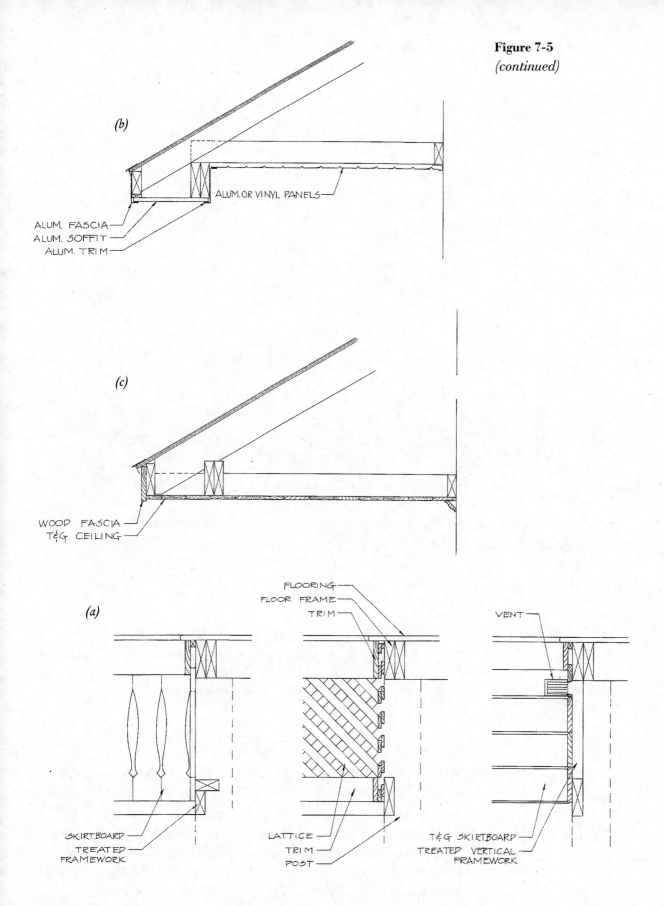

**Figure 7-6**
*Porch accessory structures. (a) Skirtings. (b) Balustrades.*

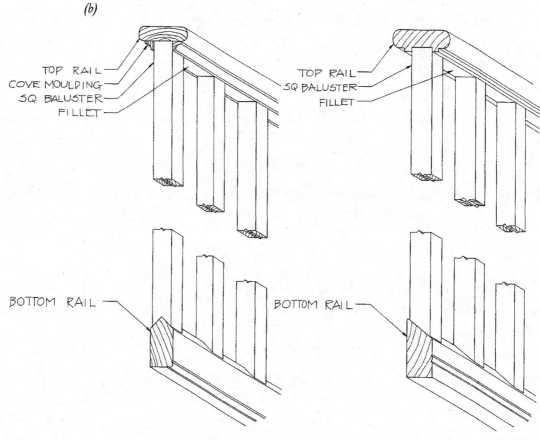

**Figure 7-6**
*(continued)*

## Deck Details

Unlike in the finer realm of porch woodwork, the decay resistance of building materials used for decks is crucial to the success and longevity of the structure. An architectural feature formerly limited to the West Coast (with its abundant supply of decay-resistant redwood), the deck has entered a new era with the advent of economical, readily available, preservative-treated lumber. Named after the planked structures on ships, the wood deck has become a widely used, fundamental feature of residential architecture, crossing many style boundaries and changing the way we approach the design of outdoor living spaces. Less formal than a porch, a deck is truly an outdoor room, making the outdoors an accessible and convenient place to spend time.

Since a deck, by definition, has no roof, its components are constantly subjected to the full forces of climatic degradation. Although some of the

# Porches and Decks

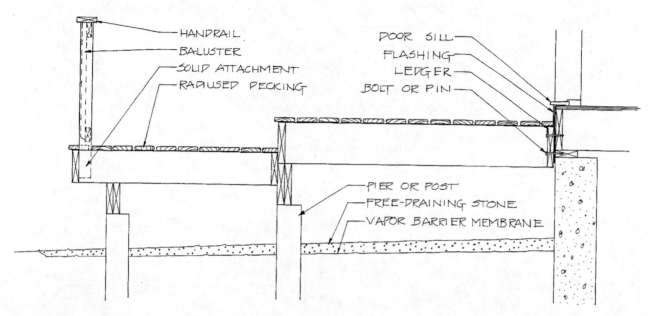

**Figure 7-7**
*The elements of deck construction.*

principles of porch design apply to the construction of decks, the exposed nature of the finished product requires other, special considerations.

Figure 7-7 illustrates the essential elements of contemporary deck construction. Construction design considerations include the following:

**1.** Since every piece of wood, both structural and nonstructural, has the potential to be repeatedly and relentlessly wetted to the point of rotting, all wood used for the construction of a deck should be of a decay-resistant type. Redwood, cedar, cypress, and pressure-treated pine are all viable choices.

**2.** In a like manner, all metal fasteners and hardware should be corrosion-resistant. Stainless steel nails, bolts, and screws are ideal, particularly for premium-grade work, seaside construction, and commercial work where the liability of failure is a significant concern. Hot-dipped galvanized steel fasteners are an excellent second choice when budget is a consideration. Electrogalvanized fasteners are not very durable, as their zinc plating is thin and it readily degrades. Most galvanized nails used in pneumatic nailers are of the electrogalvanized type and are a poor choice for deck construction.

**3.** A deck, to a greater extent than a porch floor, must be designed to rapidly and effectively shed and expel large amounts of water, in most climates. This can be accomplished by one of two methods: a slope, to let water run off, or a gapped surface. Providing a small gap between each deck board ensures that water will never collect on the deck surface. Gaps also provide for the free expansion of the individual boards upon wetting. A slope in the floor will tend to divert water away from the main building, but this is not a necessity if the deck is gapped. Complex deck designs, with odd angles, multiple levels, built-in seating, and so forth, become extremely difficult to lay out and construct when the extra variable of a slope is tossed into the mix. The decision to include a sloping floor in a deck design specification should not be an arbitrary one.

Figure 7-8 illustrates several basic deck construction details, as well as a selection of details for rails, seating, and other accessories. Detail (a) shows

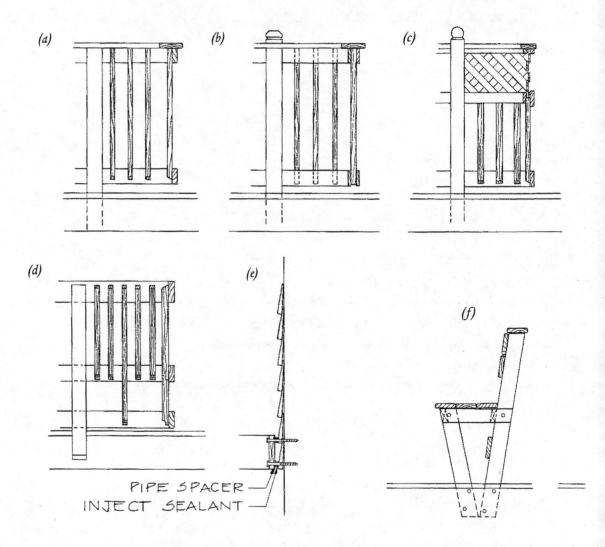

**Figure 7-8**

*Deck details. (a) A simple balustrade with an unbroken top rail. (b) A refined version with dual rails and protruding newel ends. (c) A two-tier post-to-post balustrade. (d) A two-tier balustrade with applied (rather than embedded) newel elements. (e) Ledger detail for retrofit application. (f) Bench of 2 × 6 inch frame. (g) Bench of 2 × 12 inch frame. (h) Stack-constructed bench and planters. (i) Backless bench for low decks.*

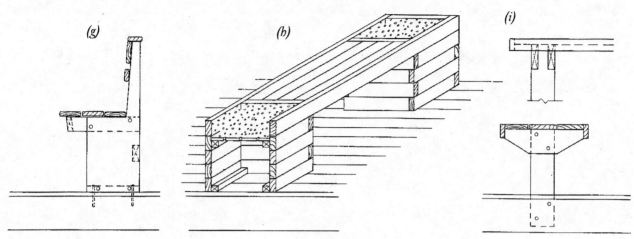

**Figure 7-8**
*(continued)*

a common and effective balustrade design, with a continuous top rail over the newel posts. The more difficult rail shown in detail (b) allows shaped posts to protrude through the top rail; the double-sided subrails provide a more finished look while allowing rapid drainage. The balustrade in detail (c) uses decorative latticework, and the one in detail (d) uses face-mounted (versus embedded) newels. Detail (e) illustrates a good method of ledger attachment when retrofitting over wood siding. Details (f) and (g) depict bench designs, which may double as railings. Detail (h) illustrates a bench of stack construction with stone-filled planters. Detail (i) shows a bench supported on upright square timbers.

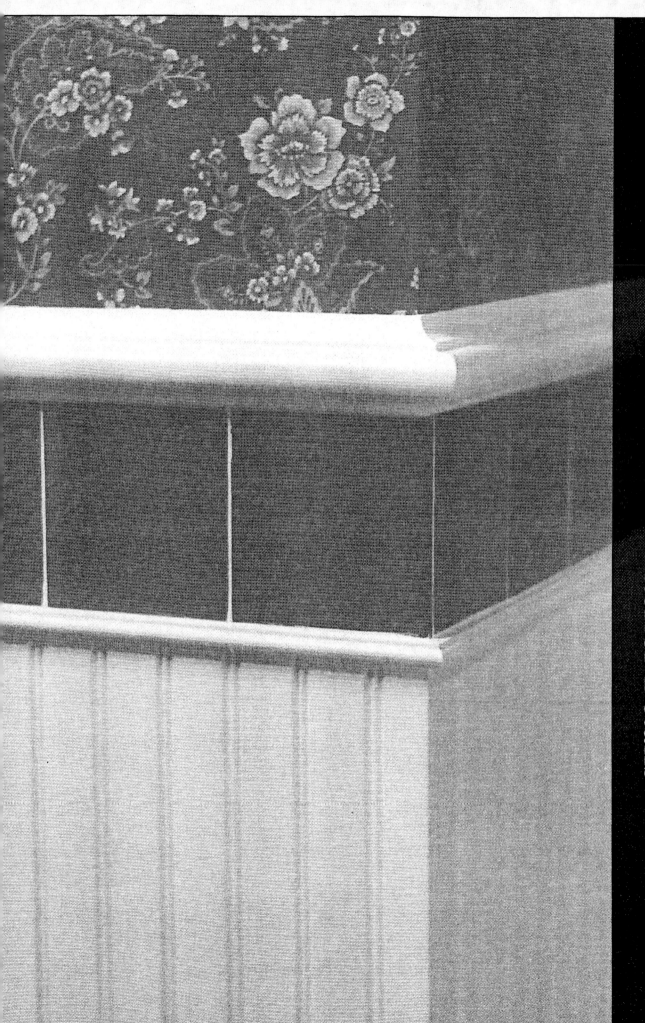

# Section Three
## Interior Details

CHAPTER 8 | # Windows

From an interior perspective, windows present a discrete field that is altogether different from the wall field with which they share a common plane. Windows dramatically alter interior spaces, adding light, air, and views; therefore they are treated here independently, rather than simply as another element of an interior wall surface.

This chapter presents a selection of interior window edge details, as well as a discussion of the problems of building and embellishing multiple windows (stacks, rows, and banks).

For a more thorough discussion of windows as fields, see Chapter 3.

## Trim Details

Figure 8-1 illustrates four distinct window trim styles, ranging from a minimalistic to an elaborate traditional approach. Each style is quite appropriate for a specific design scheme, and together they serve to identify all the prominent elements of interior window woodwork.

The trim scheme illustrated in Figure 8-1(a) is widely used in contemporary and commercial applications when a clean, uncluttered line is desired. Usually the drywall or plaster runs directly up to the window, acting as trim and jamb extension at once. No additional woodwork is used around the window—a bead of sealant closes the joint between the window and the wall. The stool, however, is subject to much greater abuse and wear than the

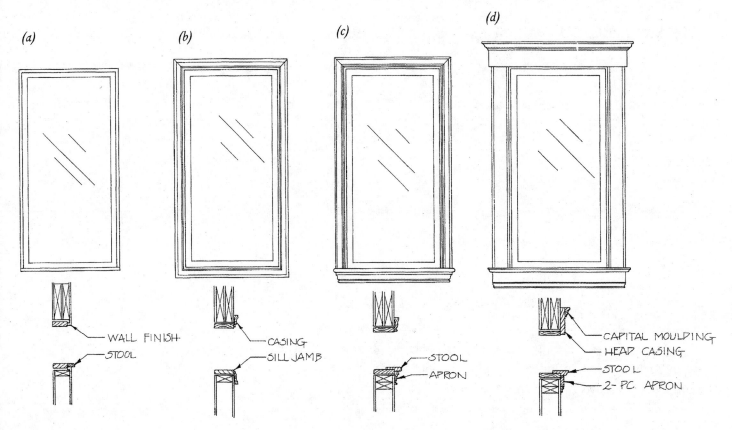

**Figure 8-1**
*Window trim assemblies. (a) Drywall finish returned to window.
(b) Asymmetrical casing, mitered all around. (c) Mitered casing on stool-and-apron trim.
(d) A classical surround.*

rest of the trim, and therefore it is usually built of a more durable material. Hardwood, solid surfacing, and marble are all common choices.

Figure 8-1(b) shows an extremely widely used trim construction. It is simple and inexpensive, and depending on the casing and finish, it can be utilized in a variety of contemporary and traditional design schemes. Unfortunately, this "picture frame" approach is so common that it often appears just that, and it is difficult to use effectively when a distinctive appearance is desired. It is best suited to contemporary designs that require a simple appearance.

Traditional design schemes are better served with a construction similar to the ones in Figures 8-1(c) and 8-1(d). The window in figure 8-1(c), the less elaborate of the two, uses mitered head casing joints and a stool with an apron as the sill trim. This design is best used with an asymmetrical casing, inverted to form the apron, with the ends returned to the wall.

The trim construction shown in Figure 8-1(d) illustrates a "high" or refined style, which draws directly on the classical orders (see Chapter 2): the apron and stool act as the plinth, the casings are the columns, and a head casing is the entablature.

A playful alternative to this classical detailing, widely used in Victorian schemes and reproductions, uses a corner block to make the turn from the side to head casings. Usually these casings are identical in this design, and the corner block is embellished with a rosette carving that often is identical in cross-section to the symmetrical casing being used.

Figure 8-2 illustrates several variations on these four basic constructions.

## Multiple Windows

When more than one window unit is used to fill an opening, additional considerations arise with respect to the windows' interior woodwork.

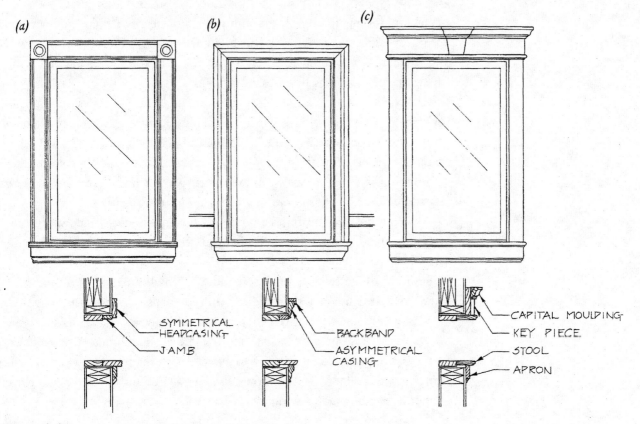

**Figure 8-2**
*Additional window trim assemblies. (a) The use of corner blocks.*
*(b) The use of a backband allows clean reception of chair rail.*
*(c) Classical surround with key piece.*

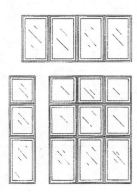

**Figure 8-3**
*The three basic multiple window arrangements (clockwise from top): row, bank, stack.*

The trim element that joins two windows together is called a *mullion*. (An example is illustrated in Figure 8-4.) The process of placing two or more windows adjacently is called mullioning or joining, and once joined, such windows are considered to be mullioned. Multiple windows occur in three basic configurations—rows, stacks, and banks—illustrated in Figure 8-3.

Two distinct joining processes can be used. The simplest and most economical is to specify that windows be factory-mullioned, or joined together by the manufacturer. The resultant multiple window unit can then simply be placed into a single opening of suitable size and site-trimmed at its outermost edges like a single window.

Structural or other considerations may preclude factory mullioning, however, in which case the windows may be site-mullioned, or joined together in a variety of manners by the installer. Each individual window must be placed into its own opening and fully trimmed at all wall-window and window-window interfaces.

Three basic factors come into play in the design of site-built multiple window assemblies: 1) structural or other constraints that affect the site's suitability for mullioning, 2) the trim scheme of nearby single windows, and 3) the specifics of the windows to be joined. A correct design will provide for a smooth, harmonic interplay of all these factors, with the end result appearing as natural as possible.

Naturally, the design possibilities of multiple windows are virtually endless; therefore, only basic examples are illustrated here, as guides. Building multiple windows on-site requires thoughtful planning and layout as well as careful craftsmanship for the finished construction to appear correct. Site-mullioning is nearly always a more expensive option; factory-mullioning should be the first consideration when a choice is available.

Figure 8-4(a) illustrates a typical factory-mullioned window. The center of the window does not lend support to the wall structure (although some window manufacturers will build combinations with structural mullions). Figure 8-4(b) illustrates a similar window, site-mullioned around two center studs. Note the use of a thin mullion trim (perhaps of ¼ inch plywood), as it must neatly join the narrow edge of the asymmetrical head casing. Figure 8-5(a) illustrates one approach to installing a site-mullioned window bank. Note the consideration given to variable frame and jamb depths and to mullion thickness. Figure 8-5(b) shows an approach to site-joining a half-circle window over a standard window.

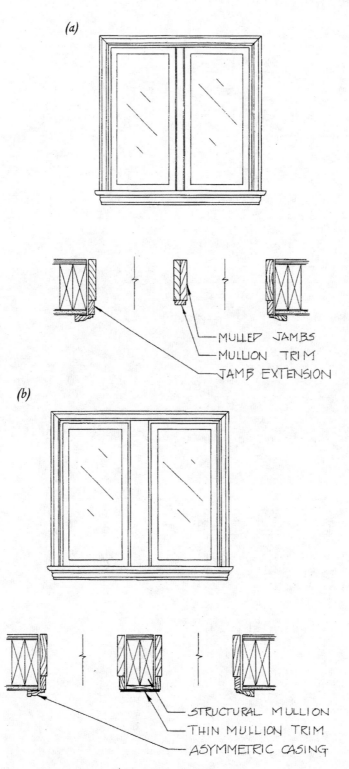

**Figure 8-4**
*Multiple window details. (a) Factory-mullioned.*
*(b) Site-mullioned, with thin mullion to abut narrow edge of head casing.*

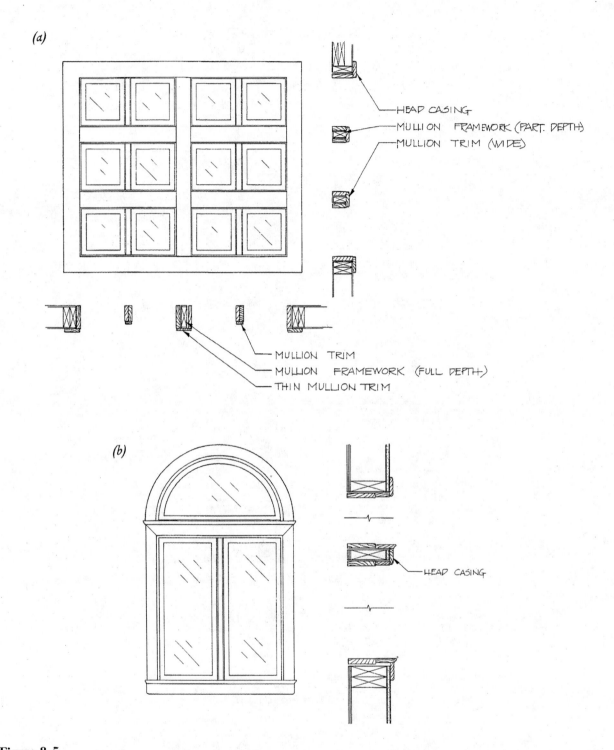

**Figure 8-5**
*Site-mullioned multiple window construction details. (a) A hierarchy of mullions: Center vertical protrudes nearly to plane of casing—its framing is full depth. Horizontal mullions with partial-depth framing. Short vertical factory mullions are also set back, and abut horizontals with thinner mullion trim. (b) The use of the head casing of a lower window as a mullion and skirt for an upper window.*

# CHAPTER 9 | Doors

Like windows, doors are discrete constructions on the interior wall plane. Doors and door trim can range from utilitarian passageways and closures to beautiful, independent design elements. Like exterior doors (see Chapter 4) the interior door's primary purpose is utility—it allows passage to and from various interior spaces, and it provides security and privacy to various degrees, as needed. As they are generally limited by the practical constraints of human scale, doors are almost always people-sized; they are only substantially larger or smaller than that for a specific design reason.

Interior doors are generally of a simpler, often lighter construction than their exterior counterparts. Absent are concerns for air and water infiltration, heat loss, and weatherability. Wood, in one form or another, is the dominant material for interior doors.

Interior doors are typically trimmed identically to interior windows in the same room; the difference being the lack of a sill/stool construction on doors, as door trim must meet the floor.

This chapter discusses the primary types of interior doors, both in terms of what they are made of (wood, core, wood by-product) and how they operate (swinging doors, bypass doors, bifold doors, pocket doors, and so on).

Some interior door edge details are presented in order to highlight the basic differences between door trim and the related trim of windows, covered in Chapter 8.

## Field Conditions

Figure 9-1 illustrates the primary types of interior doors and mounting frames, or jambs.

**Figure 9-1**
*Interior doors and frames. (a) Flush door construction. (b) Solid flush door in steel frame. (c) Hollow flush door in veneer core wood frame. (d) Moulded hard door in adjustable wood frame. (e) Built-up solid wood glazed door on heavy rabbeted wood frame.*

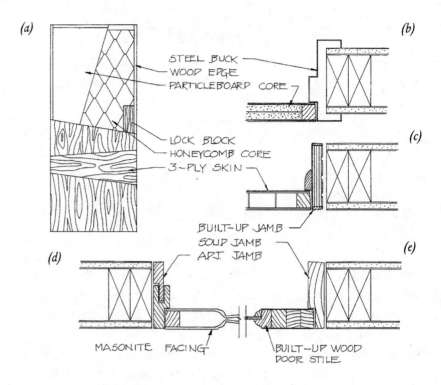

The solid-core door (detail a), mounted in a steel frame (or buck), is widely used in commercial applications. Constructed of solid particle-board between two thin plywood skins, the heavy door panel stands up well to the severe demands of a typical commercial application, and its smooth face is easily cleaned and maintained. The formed steel frame with an integral stop provides a durable mounting for the door, well suited to frequent traffic and the inevitable impact of rolling stock.

Figure 9-1 (details b through d) illustrates three types of residential interior doors, shown mounted on wood jambs.

The hollow-core door, with veneer-core plywood skins, is similar in appearance to the solid-core door shown in Figure 9-1(a). Its core, however, is composed of a lightweight honeycomb of corrugated paper or a similar filler.

The molded door is a decorated variant of a hollow-core door. The skins are made of hardboard that has been formed under heat and pressure to emulate traditional wood panels. These doors are typically manufactured with a baked-on paint primer, intended for a paint finish.

Both types of hollow-core doors are intended for light-duty use and will provide good service in low-demand residential applications. Hollow doors lack the heft or "feel" of a solid door, and they are often maligned as cheap. They *will* dent or break upon sharp impact; they have received a bad name as a result of overuse in inappropriate settings such as low-budget multiunit housing and low-budget commercial buildings, where they are destined to fail. The solid-core door is much more appropriate for such demanding applications, and it will provide better value over time.

Figure 9-1(d) illustrates a frame-and-panel door of solid wood. Paneled wood doors are available in many configurations (four-panel, five-panel, fifteen-panel, and so on) to suit a variety of design intents. Unless they are custom-built at a millworks, modern panel doors are not truly *solid* wood. They are usually a multipiece assembly of solid-wood cores, edgings, and veneers, designed to conserve dwindling supplies of expensive, clear wood stock and to provide a more dimensionally stable end product. Solid-wood doors are fine pieces of woodwork for use in period designs or anywhere that the feel and/or appearance of real wood is desired. They are commonly available in pine, fir, and oak; for other species they must generally be custom-made.

## Operating Mechanisms

Interior doors can utilize several different mechanisms to open and close. Figure 9-2 illustrates the four primary types in general use: hinged (swing), bifold, bypass, and pocket (disappearing). The hinged or swing door is, of course, the most common. Its mechanism is simple, direct, and effective. It is the easiest of all door types to operate, since the handle is placed at the point of greatest mechanical advantage, and one can use one's body weight effectively to easily open or close the heaviest of doors. Hinged doors swing clear of their opening; however, in doing so they require a large amount of space, which cannot be effectively used for anything but a door reception zone. Nonetheless, the hinged door should always be a design starting point, and it should only be abandoned when another door type is clearly superior.

Hinged doors are hung with a knuckle-and-pin hinge, called a butt, which is mortised (recessed) into both door and jamb for a flush fit and to develop the hinge's full bearing strength. The number and type of hinge is dictated by door size and weight. Light, hollow-core doors can get by with just two hinges, although three maintain better door and jamb alignment over time. Solid doors need three or four hinges, possibly of the ball-bearing type, if severe usage is expected. Butt hinges are made of plated or enameled steel, brass, or stainless steel. Ideally the hinge finish should match or complement the handle hardware finish.

Two hinged doors placed in an opposing pair (Figure 9-3) is termed a double door, or a French door if the door panels are glass panes. Double doors are used to close large openings, and can operate in one of two ways: as a stationary/operating pair, or as two independent doors. The best way to

**Figure 9-2**
*Door operation (top to bottom): swinging, folding, bypass, pocket.*

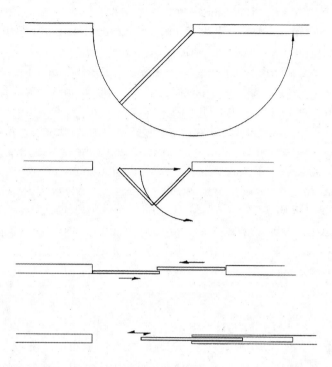

outfit a double door is to designate one panel as operating and the other as stationary. The stationary panel is fitted with specialized hardware that enables it to be secured in place, while the operating panel is fitted with a standard handle that latches into the meeting stile of the stationary panel. A stop, called an astragal, is fitted to the meeting stile of the stationary panel to keep the operating panel from overswinging.

A simpler and less costly, but somewhat imperfect, setup is to mount spring-loaded latches, such as ball catches, at the top of each door panel so they are held firmly in place at the closed position (see Figure 9-3). Dummy (nonlatching) handles are used to open and close the doors. The drawback to this method is the undue torque developed by locating the handle at such a distance from the latching point. The doors twist slightly as they open and close, possibly causing joint failure, and unless each door panel is made perfectly flat, and stays so, the alignment between the independent doors is tenuous.

An odd relative of the hinged door is the full-swing door, shown in Figure 9-3(b). Fitted with special pivots, this door opens in both directions; it stays in the closed position by means of spring or cam tension. Often used in kitchens for its hands-free operation, the full-swing door should be specified only for a distinct purpose, as it is inherently an unsafe door, especially for pets and small children.

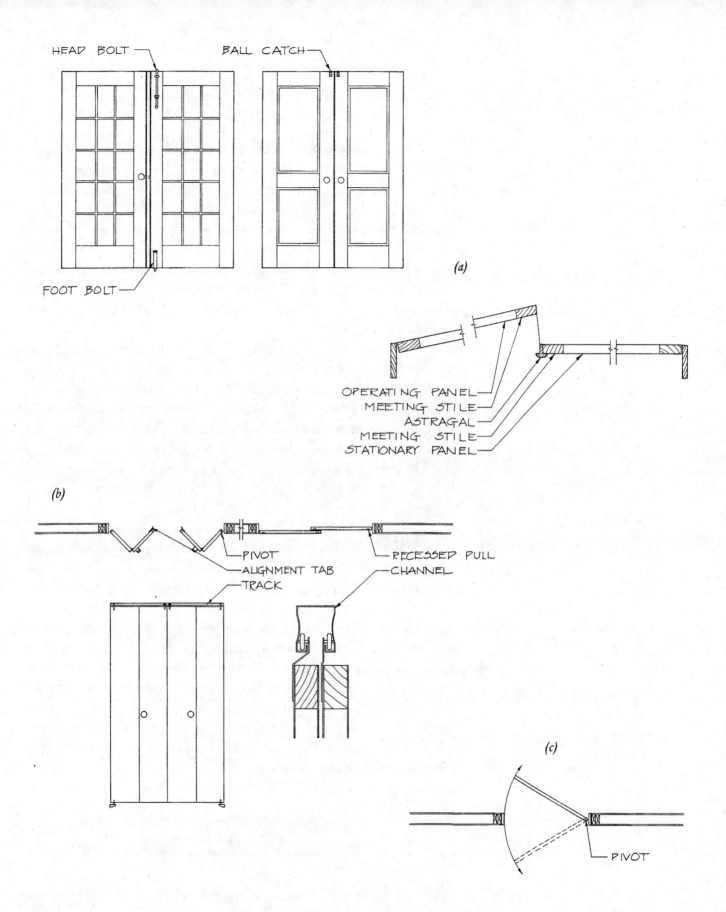

**Figure 9-3**
*Door details. (a) Two methods of configuring double doors. The pair at right are identical and require no astragal. (b) A double-acting, or full-swing door. (c) Bifold and bypass hardware configuration.*

The bifold door (and the double bifold) uses a combination of hinges, pivots, and a guide track to split and fold as it opens. Used mainly as closet doors and room dividers, their operation is a bit more complex and unreliable than that of the hinged door. Their advantage, though, is their compactness. A bifold, when opened and while opening, takes up very little room and is ideal when the space in front of the door is tight. Bifold doors should be specified with the heaviest hardware affordable; lightweight components tend to wear quickly or even fail if slightly abused. Bifold doors are fitted with either knobs or dummy handle sets, ideally placed as shown in Figure 9-3(c) for proper operation.

The bypass door, also shown in Figure 9-3(c), utilizes two or more door panels arranged to slide past each other on an overhead track mechanism. Bypass doors use up no additional floor space, but they cannot, by their nature, ever be fully opened. Like bifolds, bypass doors are well suited for closet areas in tight or obstructed quarters, but their half-opening shortcoming is often objectionable. The inner panel of a bypass pair must be fitted with fully recessed (or retractable) hardware in order to clear the outer panel when operated. Normally the outer panel is fitted with identical hardware, for consistency.

The pocket (or disappearing) door slides on an overhead track into a door-sized recess within the adjacent wall. Pocket doors are great problem solvers when space is at a premium or when a door sees infrequent use. Somewhat more difficult to operate than a hinged door, the pocket door does, however, enjoy significant popularity because it is so versatile. A pocket door must have an available wall space of approximately twice its width. Half this space becomes the pocket, the other half the opening. Several framing methods are shown in Figure 9-4. In general, the thicker wall systems are better, because they are stiffer and allow adequate room for utility runs and terminations (piping and wiring). Pocket door track hardware is available from several manufacturers, and should be chosen for its durability and suitability for the intended door type. Several latch or handle systems are available in both locking and nonlocking styles.

## Edge Details

Figures 9-5 through 9-7 illustrate several types of door trims and embellishments. Door trim varies from window trim in that it must meet the floor and intersect baseboard trim.

# Doors

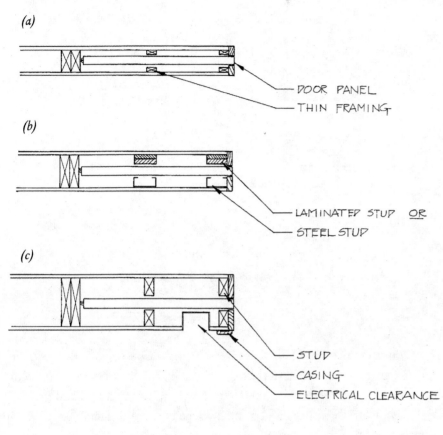

**Figure 9-4**
*Pocket door framing.
(a) Small wood or steel-wrapped wood studs provide clearance for door, but do not resist lateral deflection well. (b) Nominal two-inch studs of laminated wood or plywood or of light-gauge steel provide good stiffness in a minimum of space. Solid wood framing, installed flat, will work in this configuration, too, but may cause problems with alignment if warping occurs. (c) Wood studs, aligned typically, provide excellent stiffness and generous space for utilities.*

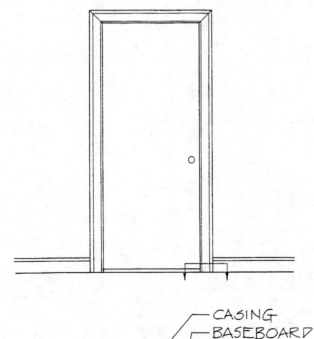

**Figure 9-5**
*Asymmetrical casing.*

**Figure 9-6**
*Symmetrical casing with backband to cleanly receive wainscot.*

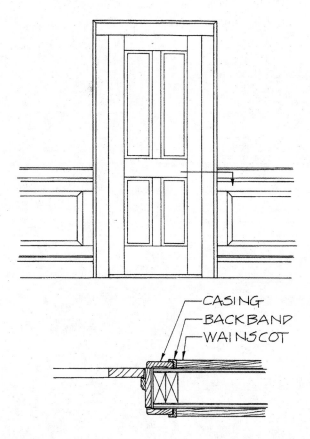

An asymmetrical casing (Figure 9-5) should be slightly thicker at its outer edge than the thickest part of any adjoining baseboard or (ideally) chair rail. This assures a clean intersection with fully concealed end grain.

Symmetrical casing can be fitted with a backband (Figure 9-6), which essentially creates an asymmetrical profile for a proper baseboard joint; or a thicker plinth block can receive both the casing and the baseboard with elegant simplicity (Figure 9-7).

**Figure 9-7**
*A traditional casing with plinth blocks to cleanly receive baseboard.*

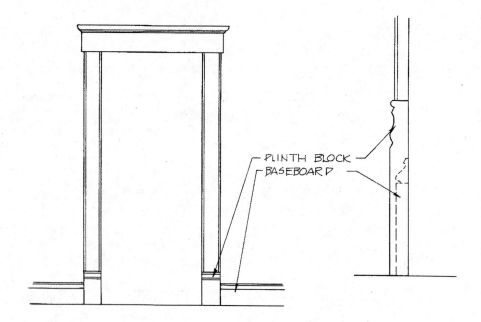

# CHAPTER 10 | Ceilings

This chapter and the two that follow, on walls and floors, deal with the three primary *surfaces* of a building's interior. Various types of woodwork can define, decorate, or simply delineate these surfaces; the opportunities for creative design are manifold.

Typically the ceiling surface, whether level, vaulted, or otherwise constructed, is left with a smooth or textured drywall or plaster finish that meets, unceremoniously, the similar finish of the wall at a simple inside corner. The use of woodwork as a finish material is unnecessary.

In higher-end work, or as part of a renovation or restoration project, the ceiling can become much more than a plain, flat surface for paint. The ceiling may itself be constructed of wood strips or planks, or structural or false beams may create a protruding field condition that can be treated with some form of woodwork. The edges where the ceiling meets the wall or the vertical side of a beam are excellent spaces for embellishment, as these joints may be defined by a single or multiple crown or similar moulding.

This chapter describes and illustrates a variety of construction details for the architectural woodwork of ceiling surfaces. Several wood ceilings are presented, followed by a selection of beam details and edge moulding assemblies (crowns).

## Wood Ceilings

There are two basic types of ceilings made primarily of wood materials, as illustrated in Figure 10-1. The plank ceiling, where wide boards of either structural

# CHAPTER 10

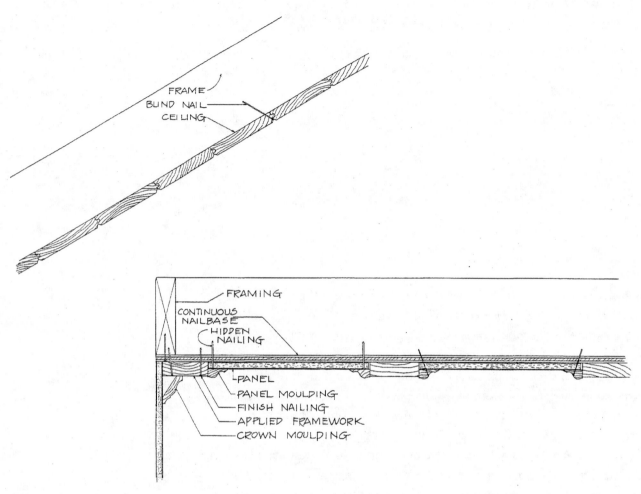

**Figure 10-1**
*Wood ceilings: plank construction (top) and paneled.*

or nonstructural origin interlock to form the ceiling surface, is quite common in contemporary and commercial design. (A more refined version, also described in Chapter 7, is the strip ceiling, made of narrower boards, often milled with a beaded profile.) Second is the paneled ceiling, which can range from a simple construction of plywood with applied edge mouldings to an elaborate system of raised or recessed panels and accompanying mouldings. Several additional examples of various wood ceiling details are illustrated in Figure 10-2.

The application of the wood itself is generally quite straightforward. The difficulty with much wood ceiling work is providing an adequate base for attachment. The framework or application base should be reasonably flat and smooth, and it must provide a good base for proper fastening of material. Often the addition of furring or other secondary framing or blocking is necessary.

# Ceilings

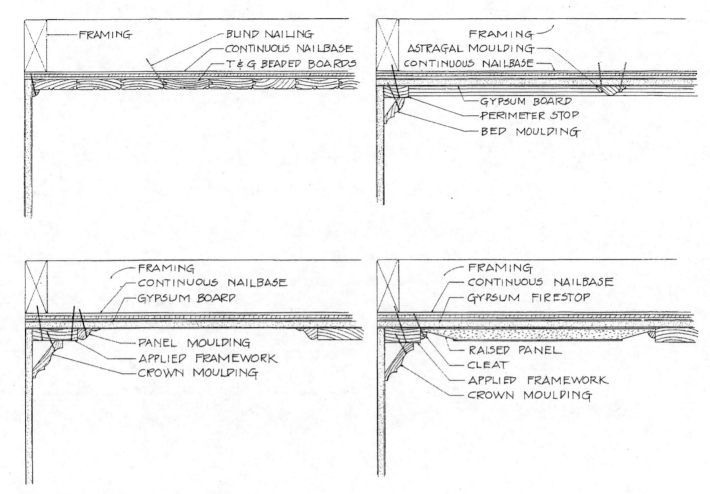

**Figure 10-2**
*Additional wood ceiling details.*

## Beams

Beams at the ceiling surface are of two distinct varieties: structural (necessary) and decorative (unnecessary). Often the two are used in combination to disguise or enhance visible structural elements.

Figure 10-3 illustrates several beam field conditions, ranging from a simple wrapped structural beam to an elaborate network of interlocking structural and decorative beams called a coffered ceiling. Note the use of a continuous nail-base sheathing as an economical and foolproof method of providing proper attachment of the various woodwork components.

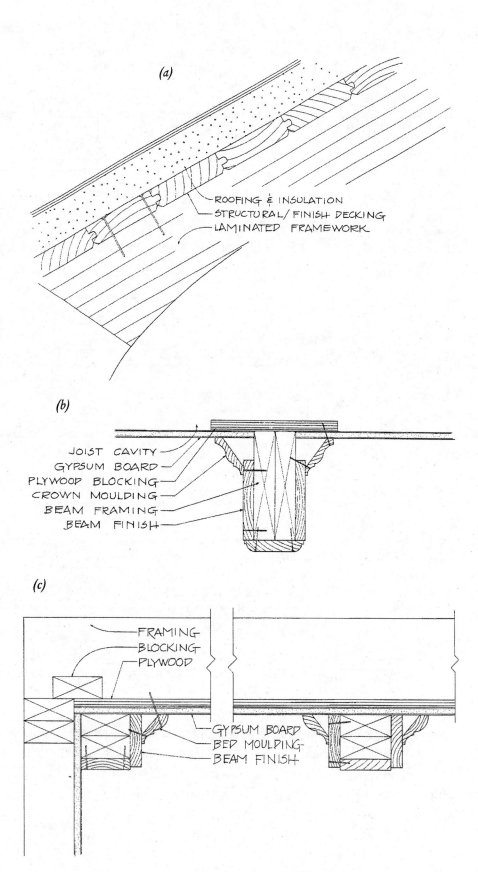

**Figure 10-3**
*Beam details.
(a) Structural beam supports ceiling surface directly. (b) Cased structural beam supports framework above.
(c) False (built-up) beams of a coffered ceiling.*

# Edge Conditions

Figures 10-4 and 10-5 illustrate built-up crown moulding profiles, of wood and wood mouldings. The scale and alignment of crown moulding should reflect the size and height of the room in which it is installed. Room height in particular affects perspective, as shown in Figure 10-4(b). Low ceilings (less than about nine feet) cause the crown to be viewed more from the side than from the bottom, so the profile should ideally extend downward more than outward. Most stock crown profiles are milled this way. High ceilings change the perspective, and the crown should ideally be larger and should project out onto the ceiling at least as much as it drops onto the wall. The use of built-up crowns makes these adjustments a fairly easy task.

Figure 10-5 illustrates the use of a crown construction to conceal indirect lighting on a vaulted ceiling.

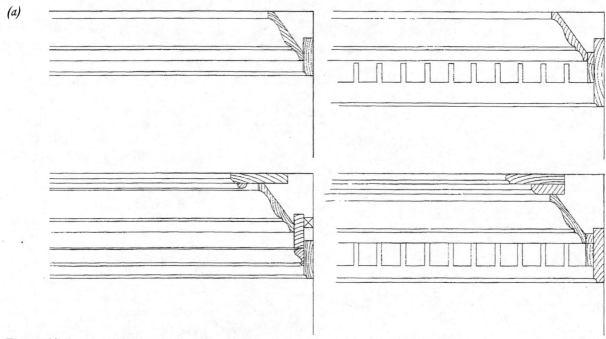

**Figure 10-4**
*Built-up crown mouldings. (a) two-, three-, seven-, and five-piece crown profiles, with various height and projection dimensions.*
*(b) Overall crown profile should consider the perspective of the viewer. Higher ceilings generally require a greater relative projection.*
*(c) A built-up crown retrofitted to an existing ceiling.*

# CHAPTER 10

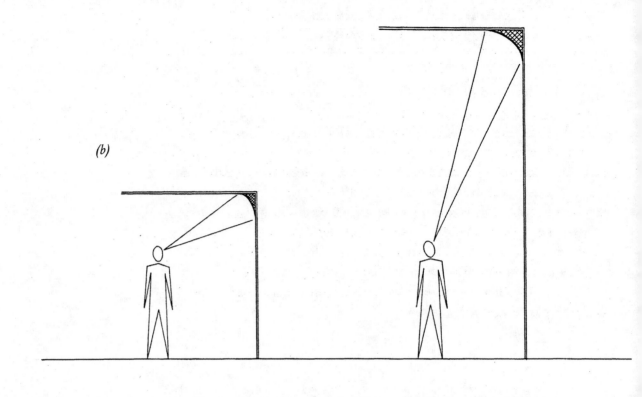

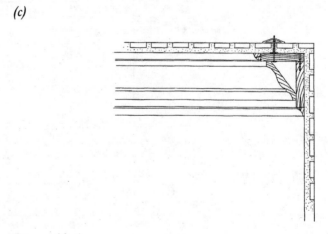

**Figure 10-4**
*(continued)*

## Synthetics

A wide variety of period crown mouldings are also available from several manufacturers of synthetic moulding—generally a form of moulded foam plastic with workability similar to wood. These mouldings are generally

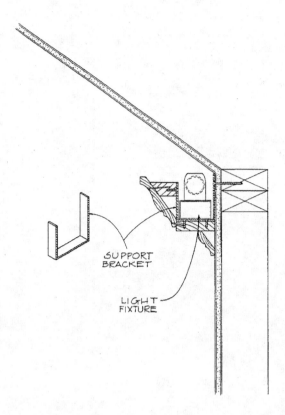

**Figure 10-5**
*Concealed lighting integrated into a crown profile.*

intended for a painted finish, and they may even come with special inside and outside corner pieces as substitutes for complicated miter and cope joints.

The main drawback to these otherwise flawless mouldings is the prominent joint where two sections come together (see Figure 10-6). Unlike a built-up wood moulding with obscured, staggered joints, large synthetics must use large joints; these are often difficult to conceal, especially if the wall and/or ceiling surface or framing has any substantial surface imperfections. Also, the available lengths of synthetic mouldings are limited, creating more joints in a large room than would be necessary with wood construction. Nonetheless, if specified and installed carefully and for an appropriate application, synthetic mouldings are able to provide very detailed and elaborate profiles at reasonable cost.

CHAPTER 10

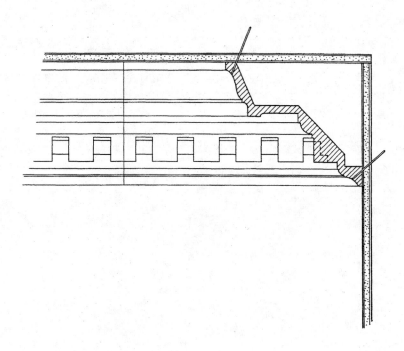

**Figure 10-6**
*A typical synthetic crown moulding, with prominent joint.*

# CHAPTER 11 | Walls

## Introduction

The interior wall field is generally the most noticeable surface in a building, as our eyes tend to see it clearly without effort (unlike the ceiling or floor, which are observed only by looking up or down). In terms of wear, it is subjected to more abuse than a ceiling and less than the adjacent trodden floor surface. Generally it is the lowermost portions that are the most prone to damage—hence the common use of protective design elements such as chair rails and wainscots or, at a minimum, baseboards.

From a historical perspective, a wall can be treated like a column of a classical order, with baseboards acting as the plinth, wainscot panels as pedestals, crown moulding as capital, and the wall itself, with any included woodwork, as the column shaft (see Figure 11-1).

Figures 11-2 and 11-3 illustrate a series of wall woodwork systems of differing complexity. These figures illustrate the parts of common wall woodwork, as well as several options for effective installation.

Figure 11-2 illustrates four wainscoting constructions. The one shown in detail (a) uses the wall surface (or a nail-base panel flush with the upper wall plane) as the wainscot's recessed panel. The lower framework also acts as a baseboard. This design, although of fairly simple construction, offers an excellent traditional appearance for the money. The more elaborate and three-dimensional design of detail (b) utilizes a separate baseboard and a protruding rail cap. This construction requires thick casings or casing backbands (see Chapter 9) to neatly terminate these elements at wall openings.

# CHAPTER 11

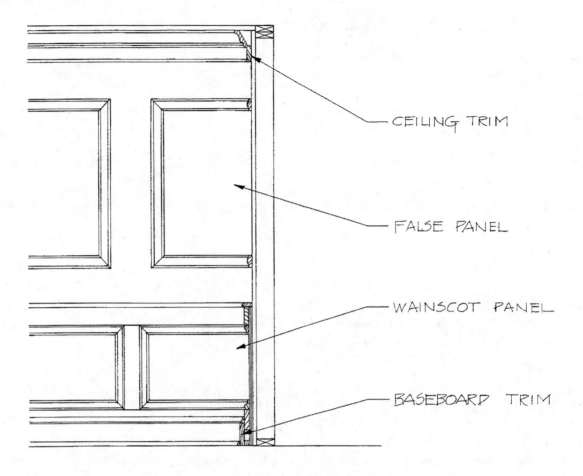

**Figure 11-1**
*A classically derived wall.*

The wall in detail (c) utilizes applied raised panels constructed for easy removal should they be damaged. The use of traditional frame-and-panel construction, with the panel locked into a groove in the frame, is unnecessary in one-sided work like this. Detail (d) shows a vertical wainscot attached to horizontal strapping.

Figure 11-3(a) illustrates a full applied framework. Panel moulding surrounds the upper wall surface panels, while the lower panels are raised. Figure 11-3(b) shows a faux paneled wall, which uses simple and inexpensive frames of applied moulding to define "panels" on the wall surface. Each of these designs benefits from forethought at the rough framing stage, when properly positioned blocking can easily be added.

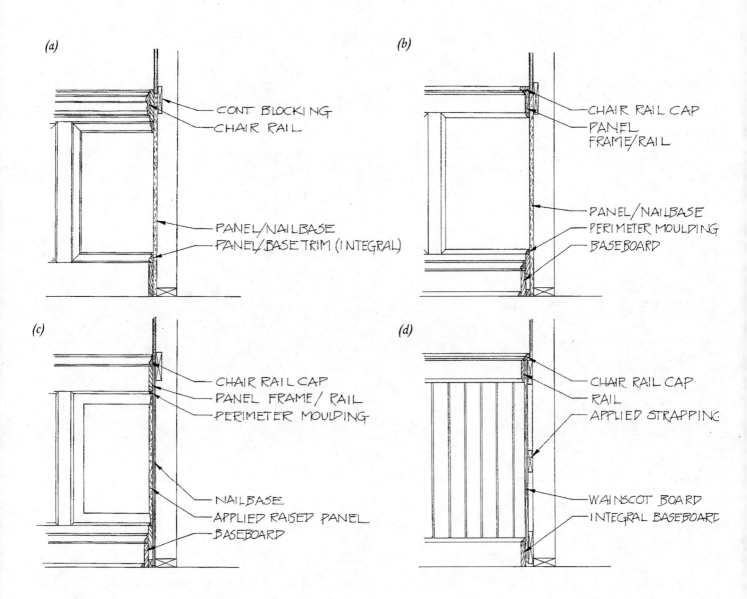

**Figure 11-2**
*Wainscot details. Note the use of continuous nailbase sheet goods, as well as continuous blocking, which helps to align errant studs. Since wall trim is one-sided, panel construction is simpler than for (two-sided) cabinet work.*

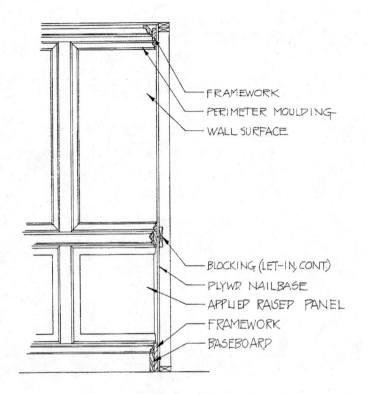 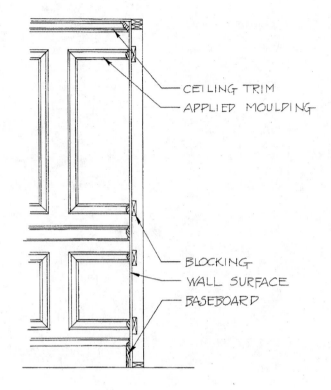

**Figure 11-3**
*Paneled wall details. Simple applied mouldings (right) can be attractive, but they usually lack the depth and authenticity of true paneled work (left).*

# CHAPTER 12 | Floors

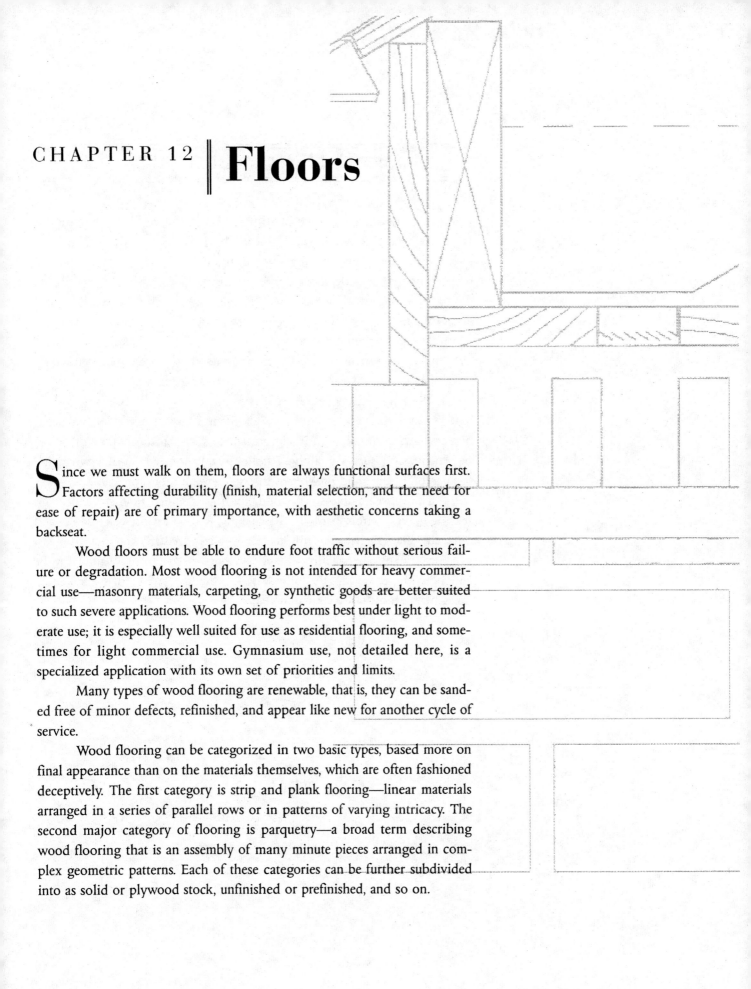

Since we must walk on them, floors are always functional surfaces first. Factors affecting durability (finish, material selection, and the need for ease of repair) are of primary importance, with aesthetic concerns taking a backseat.

Wood floors must be able to endure foot traffic without serious failure or degradation. Most wood flooring is not intended for heavy commercial use—masonry materials, carpeting, or synthetic goods are better suited to such severe applications. Wood flooring performs best under light to moderate use; it is especially well suited for use as residential flooring, and sometimes for light commercial use. Gymnasium use, not detailed here, is a specialized application with its own set of priorities and limits.

Many types of wood flooring are renewable, that is, they can be sanded free of minor defects, refinished, and appear like new for another cycle of service.

Wood flooring can be categorized in two basic types, based more on final appearance than on the materials themselves, which are often fashioned deceptively. The first category is strip and plank flooring—linear materials arranged in a series of parallel rows or in patterns of varying intricacy. The second major category of flooring is parquetry—a broad term describing wood flooring that is an assembly of many minute pieces arranged in complex geometric patterns. Each of these categories can be further subdivided into as solid or plywood stock, unfinished or prefinished, and so on.

# Solid Wood Strip and Plank Flooring

Figure 12-1 illustrates varieties of strip and plank flooring. The defining difference between the two is width—strip flooring is generally less than three inches wide, and plank flooring three inches or wider. Note that the mating faces of the tongue-and-groove flooring are beveled slightly, to ensure that the top edge meets as tightly as possible. The relief cuts and channels break the surface of the boards' bottom side, helping to reduce the potential for cupping.

The most common method of attaching of these flooring materials is by blind nailing, making use of the tongue-and-groove interlock to fully secure both edges of every board. Various nailing machines can be used to install flooring—these tools are designed to draw the flooring tightly together and nail it at once. Many excellent texts are available that detail proper flooring installation. It's important to spend adequate time on flooring layout to ensure a straight, parallel installation, particularly when the installation is large and progresses through several rooms or spaces.

The most durable solid wood flooring is milled from stock of high specific gravity—from hardwood, generally. Red and white oak, maple, and birch all make excellent flooring materials—they are hard, and they have fair to good dimensional stability. Beautiful floors of cherry and walnut can be laid; however, these slightly softer woods are best reserved for special situations, where foot traffic is guaranteed to be light.

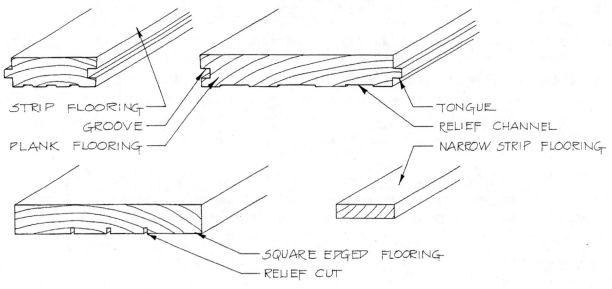

**Figure 12-1**
*Strip and plank flooring.*

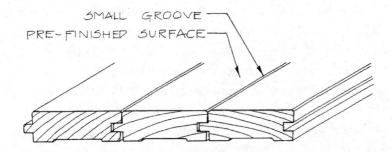

Figure 12-2
*Prefinished flooring.*

When minor denting is not likely or will not be considered a defect, then softwood flooring is a possibility. Softwood flooring is described more fully in Chapter 7. The moisture-sensitive southern yellow pine, not recommended for exterior porch flooring, is well suited for interior applications where it is likely to hold its dimensions. It is quite hard by softwood standards, and its contrasting grain is visually striking.

## Site-Finished versus Prefinished Flooring

Most styles and patterns of wood flooring (strip, plank, veneer, parquet tile, and so on) are available with preapplied stains and finishes and need only be fastened in place to complete their installation. In fact, with many engineered laminar wood floorings there is not a choice—these proprietary products are available as prefinished, ready-to-use materials only. In the case of solid wood strip or plank flooring, however, there is a choice, and in most cases the best option is to site-finish the flooring.

Site-finishing is a two-step process: the flooring surface is first sanded smooth and level, and then a series of finish coatings are applied to protect and beautify it.

All installed solid wood flooring retains minute variations in thickness due to milling tolerances and substrate variables. These variations are smoothed out during the sanding step, and once finished the floor surface is smooth and virtually free of defects.

Prefinished flooring, while its finish is generally of impeccable quality, cannot benefit from the crucial sanding and leveling process. The manufacturers of these flooring materials have compensated for this limitation by milling a slight bevel or radius along the edge of each prefinished floorboard (see Figure 12-2). The resultant groove between each board of the installed floor attempts to disguise or hide thickness variations and gaps by providing even, obvious break-lines. The bevel or radius is also necessary to prevent splintering and to provide a "ramp" of sorts to guide the foot up and onto the board (to prevent tripping).

Unfortunately, prefinished flooring products are manufactured to satisfy the convenience of the builder or installer (timing, ease of installation, and so on) rather than the long-term needs of the building (ease of repair, ability to renew or refinish). Therefore they are not always the best long-term choice. Prefinished flooring, while initially beautiful and quite easy to install, should be regarded as a temporary floor covering such as carpeting, which, when worn out, is simply replaced. Worn prefinished flooring cannot easily be sanded and renewed, due to the presence of the grooves between each board. Refinishing such a floor is a tedious, time-consuming process; each board must be separately stripped clean and a new finish applied. The results are often mixed, depending on the skill and patience of the unfortunate refinisher.

## Installation Considerations

Solid wood strip and plank flooring must be handled and laid with care in order to ensure a successful, lasting installation. Interior wood flooring, while not generally subjected to the same degree of moisture-related stress as porch flooring, is certainly prone to some moisture-related problems during and after installation. Furthermore, the consequences of moisture-related wood movement tend to be more noticeable indoors, as interior woodwork is nearly always installed to closer tolerances than exterior woodwork.

Flooring (as well as other fine millwork) should not be delivered to a job site until all of the "high moisture" work is complete, including concrete, drywall, and plaster work as well as water-based painting. Each of these operations release huge volumes of water vapor into the surrounding atmosphere, and any dry, porous material in the vicinity will, by diffusion, tend to absorb some of it. Ideally the framing and other wood at the site should be at or near equilibrium moisture content; and if central heat or air conditioning is available (and in season), then it should, by all means, be operating from the time of delivery until the final installation and finishing is complete.

In temperate climates, spring and early summer are particularly poor times of the year to install wood flooring, as the ambient relative humidity is likely to be very high and it is difficult to keep the moisture content of the flooring low enough to ensure a lasting "tightness" through the dry heating season. Thus floors installed during humid weather will likely develop some gaps in winter. This defect, although not completely preventable, can be minimized by the use of a dehumidifier during acclimatization and installation, and by including a permanent humidifier as part of the building's central heating system. The object of these precautions is to keep the flooring (and other woodwork) at a fairly constant moisture content throughout the year, avoiding large fluctuations in humidity, which will cause correspondingly large fluctuations in the dimensions of the wood.

Flooring should be installed with a gap at both edges (see Figure 12-3) to allow for unrestricted seasonal moisture-related expansion. The size of the gap will depend on the size of the floor and the relative moisture content of the flooring at the time of installation. If dry flooring is installed during the dry season, it can be expected to swell noticeably in a direction perpendicular to its grain during the humid months. For average floors of twenty feet wide or less, a ½–inch gap at either end is enough. Wider floors need a wider gap and may also need additional trim to conceal it, such as a base shoe moulding. For example, large maple floors in gymnasiums often need an expansion space of several inches. A gap should also be allowed at the ends of the flooring's long direction (at the end grain), but not nearly so large a gap as at the edges. A base shoe moulding used with any solid wood flooring should never be nailed to the flooring itself, lest it be carried out of place by the incessant seasonal movement of the floorboards.

Solid wood flooring less than three or four inches in width is adequately secured by blind-nailing through the tongue, as shown in Figure 12-4. The length, type, frequency, and location of the nails is dictated by the framing and subflooring conditions. Generally, thick, sound plywood sheathings provide a good nail base for flooring. Thin subfloors require that nails penetrate framing members as well as sheathing. Deformed shank nails should always be used to minimize potential squeaks.

Solid wood flooring greater than about four inches in width may not be able to be adequately secured by blind nailing, as these wider floorings tend to cup or warp more noticeably, and blind nailing alone may not be able to resist such movement.

Several options are available to ensure a successful wide-plank flooring installation. The simplest and most obvious is to incorporate some form of face (visible) nailing or attachment to supplement the blind nailing. Figure 12-5 illustrates several scenarios. If face fastening is used, the fasteners can be concealed with plugs or fillers; conversely, they can be accentuated with plugs of contrasting species or by the use of decorative fasteners, such as old-fashioned cut nails. Extremely wide (eight inches or greater) wood flooring should not be immovably secured at its extreme edges, lest it crack upon moving against the tight restriction. Rather, very wide plank flooring should be secured with fasteners set in a bit, as illustrated in Figure 12-6. In a like manner, square-edged flooring *must* be face-fastened, using any of these options. Also, whenever a wide floorboard is screwed in two locations (see

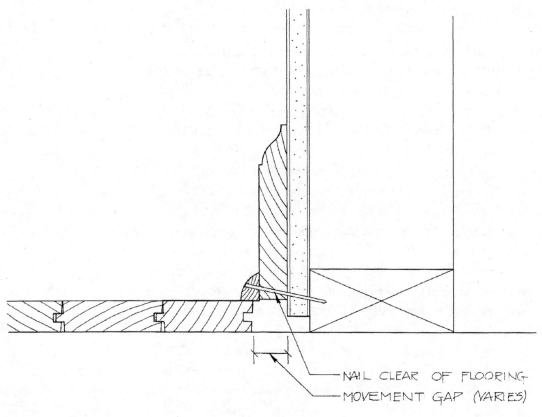

**Figure 12-3**
*Floor-wall intersection.*

Figure 12-6), one screw should be through an oversized pilot hole to allow some freedom of movement.

Another, less common, method to keep plank flooring flat is to minimize variations in moisture stress by completely finishing all sides of every board, as recommended in Chapter 7 for porch flooring. Prior to installation, the flooring is coated once with a layer of the intended final finish, particularly on the wide

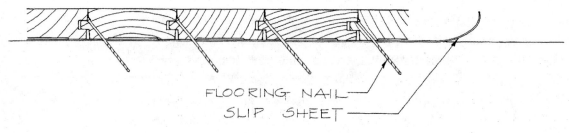

**Figure 12-4**
*Blind nailing.*

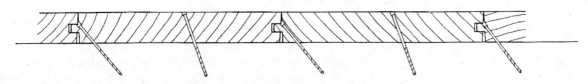

**Figure 12-5**
*Blind and exposed nailing for very wide planks.*

backside of the boards. After installation, sanding and finishing should progress rapidly. The end result will be a reduction in the floor's tendency to cup or warp, as each face will gain and lose moisture at a more constant or even rate. Blind nailing alone will be more successful if this measure is taken.

Finally, excess movement can also be minimized with the use of quartersawn stock, which, as described in Chapter 1, is much more stable across its width than the more common plain-sawn material. Using very stable stock may allow the use of blind nailing alone, even for fairly wide planks.

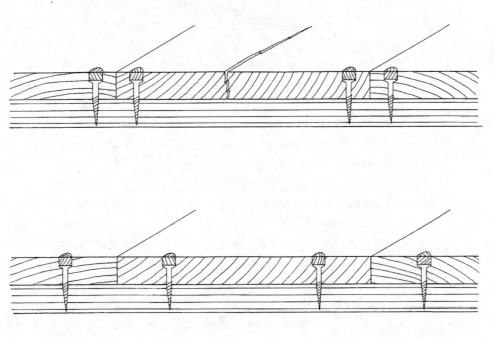

**Figure 12-6**
*Fastening wide boards. Fasteners placed inward to allow minor movement (bottom) and to avoid cracking from restraint (as in top).*

## Laminar Flooring

Using the inherent stability of plywood technology, flooring composed of a hardwood veneer laminated to a base of wood plies provides an alternative to solid wood flooring and its idiosyncrasies. Always manufactured prefinished, these laminar floorings are designed for quick, trouble-free one-step installation, either by blind nailing or gluing. Usually thinner than standard flooring, laminar flooring requires a substantial substrate for proper support.

Since their finished surface is a very thin layer of wood, these materials are not intended for heavy use, nor can they be easily restored if worn or damaged. They are well suited for light-duty use or for temporary applications such as display areas of retail interiors.

Since the integrity of these flooring materials depends on a glued layer of wood, an accidental soaking could cause irreparable delamination (whereas a solid wood floor may swell a bit and return to its previous state unchanged).

## Parquet Flooring

While strip and plank flooring are the common workhorses of the wood flooring realm, parquet flooring and the art of parquetry allow a higher level of creative expression with their intricate, often contrasting, geometric patterns.

Originally the work of skilled artisans who labored long, tedious hours to produce a bit of flooring finery, today parquet flooring is generally applied as manufactured tiles or sections, each constructed with a mosaic of small wood pieces or wood veneers. These easily installed tiles create a fair representation of traditional inlaid parquetry at a fraction of the expense.

Usually installed with a troweled adhesive, parquetry tiles (or squares) are laid out much as ceramic tile or other modular materials; they are excellent for use as border strips and contrasting zones, as well as for large-scale patterns. Several examples are shown in Figure 12-7.

If parquet and strip flooring are to be used together, an allowance may need to be made for differences in thickness—often the parquetry needs additional build-up of its substrate to bring it flush with the thicker strip material.

Regardless of the method of application, the designer must be aware that parquet patterns can be very striking in appearance and should not be overused, lest the floor appear busy. Parquetry is best suited for accent floor-

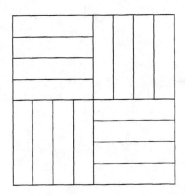 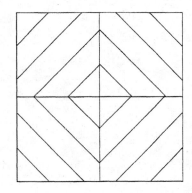
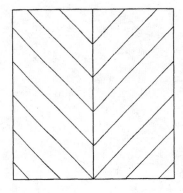 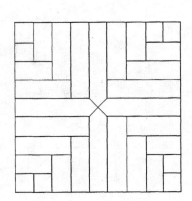

**Figure 12-7**
*Parquet patterns.*

ing, as in an entryway, where its pattern mirrors or complements a distinctive ceiling, window, or chandelier. The arbitrary use of a highly contrasting pattern throughout a large space may be tiresome, as it always visually competes with other decorative elements such as wall coverings or window treatments.

## Transitions

Figure 12-8 illustrates several strategies for dealing with floor transitions. Often a wood floor must meet a floor made of dissimilar materials. If the finished floor thicknesses are equal, then the floors can often butt together unceremoniously. Most carpeting can be joined this way to wood flooring, although the use of a terminal band across the ends of the floorboards can give a neater appearance, as in detail (a). Ceramic tile installed on a wooden deck typically requires a substantial underlayment, causing the tile surface to

**Figure 12-8**
*Flooring transitions. (a) Wood to carpeting. (b) Wood to ceramic tile. (c) Wood to tile (bath areas). (d) Wood to thin sheet vinyl. (e) An applied wood threshold, not recommended. (f) A balcony edge.*

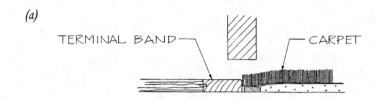
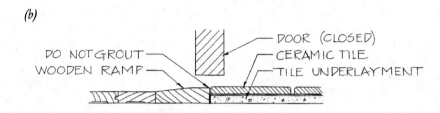
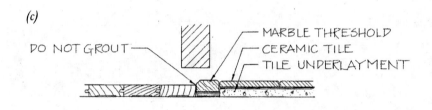
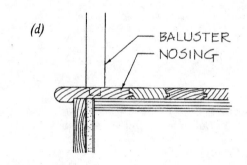
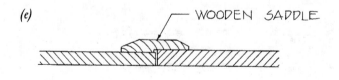
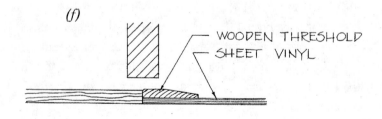

protrude above any adjacent wood flooring. If the thicknesses are close, a strip of heavy beveled brass can be screwed across the joint to hide it; detail (b) illustrates the use of a beveled wood transition threshold embedded in the floor. In bath areas, a marble threshold is commonly used, aligned as in detail (c). Very thin floorings, such as sheet vinyl, are best installed over thick underlayments to bring them close to being flush with adjacent wood flooring. If a thin underlayment must be used, then a wide gradual ramp threshold is a good solution, as shown in detail (d). The applied wood threshold shown in detail (e) should always be avoided. It is crude and usually demonstrates a lack of planning. When flooring ends at a balcony or stairwell, the common nosing seen in detail (f) makes an excellent termination.

# CHAPTER 13 | Fireplace Woodwork

A fireplace, like a window or a door, is an opening within a wall field. It must be constructed according to the general design scheme of the room or space it occupies. Woodwork applied around the opening of a fireplace can create a variety of overtones, ranging from starkly plain to classically ornamental.

Fireplace woodwork is often used to create a focal point in a room, and as a result it is often more elaborate or intricate than nearby window, door, or accessory trim. Ideally, though, fireplace woodwork should not stray from the central design concept of the rest of the building or space, lest it appear to be an add-on or afterthought. For this reason, prebuilt fireplace trim assemblies may not always be the best choice when a particular design is sought.

Aside from concerns of woodwork continuity, fireplace woodwork is controlled to a certain extent by the design of the fireplace itself and the codes or regulations that prescribe clearances to combustible materials.

This chapter presents design considerations and a range of fireplace woodwork solutions for typical varieties of fireplaces.

## Fireplace Constructions

There are two basic types of fireplaces—masonry and manufactured. The type of fireplace dictates certain practical woodwork design and construction considerations, such as attachment, clearances, and proportion.

# CHAPTER 13

**Figure 13-1**

*Vertical cross-section of a typical masonry fireplace, with a projecting firebox.*

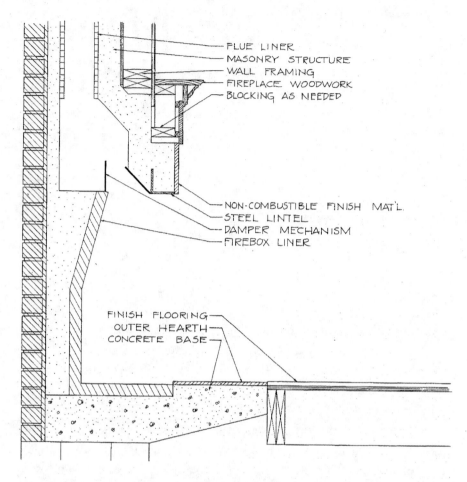

Dating to antiquity, the masonry fireplace is a massive construction of brick, CMU (concrete masonry units, commonly termed concrete block), and/or other related materials bonded with cementitious mortar; it exhausts combustion products through a chimney of like materials. Masonry fireplaces require structural footings and foundations for adequate support. A typical example is illustrated in Figure 13-1.

A manufactured fireplace, generally constructed of heavy-gauge insulated sheet metal with a light masonry firebox liner, is designed to reduce the labor costs associated with traditional fireplace construction. It requires no structural foundation and can be placed in immediate proximity to combustible materials (within certain limits). It requires an inexpensive, easily installed metal chimney, often of two or three layers. A typical manufactured fireplace is illustrated in Figure 13-2.

Other, less common fireplace configurations exist, such as the two-sided, the peninsula, and the island fireplace. Woodwork for these fireplaces often requires special design considerations, although most of the details and recommendations for in-wall fireplaces discussed below are applicable to them as well.

# Fireplace Woodwork

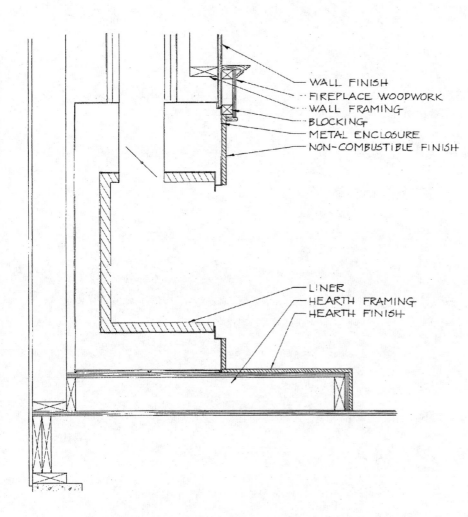

**Figure 13-2**
*Vertical cross-section of a typical manufactured fireplace, with a flush (nonprojecting) firebox.*

## Design Considerations

A good fireplace woodwork design, like most worthwhile endeavors, requires a certain amount of planning and coordination so that the fireplace and its surrounding structure is built and installed in such a way that the intended woodwork may be properly installed.

In most instances, the fireplace proper is built or installed prior to the application of wall finish materials—it is part of the rough construction, usually accomplished in conjunction with or immediately after the rough-framing phase of building. Wall finish materials, typically drywall or plaster, are then brought up to the fireplace boundary, creating a joint, which is concealed and embellished with additional finish materials: brick, tile, marble, and/or woodwork.

# CHAPTER 13

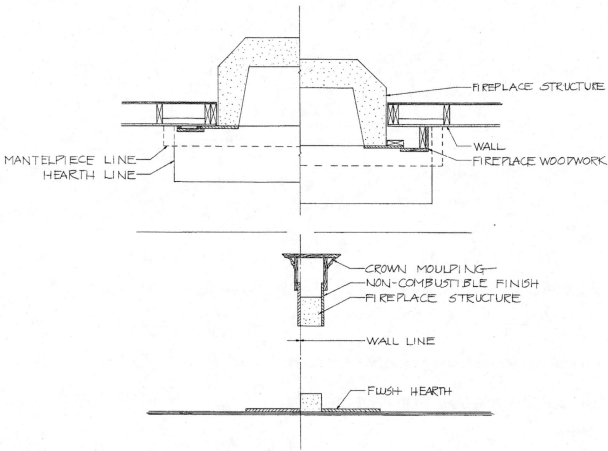

**Figure 13-3**
*Effect of firebox projection on the depth of woodwork.*

Four variables affecting the detailing of the finish woodwork are determined at this rough phase of fireplace construction: the projection of the firebox, the height and configuration of the hearth, the size of the fireplace opening, and provisions for attachment of finish components.

Figure 13-3 illustrates the effect of fireplace projection. A flush, or zero, projection necessitates a trim scheme of low relief—essentially an applied assembly of low-relief trim components. In contrast, a projecting firebox causes a like projection in the woodwork, creating high relief or depth of detail. The woodwork becomes much more three-dimensional. Projection cannot be excessive, however, as it is limited by the internal construction of the firebox and chimney, which must exit upward. Generally, a projection of three to six inches is adequate to develop the depth of detail necessary for correct period designs.

# Fireplace Woodwork

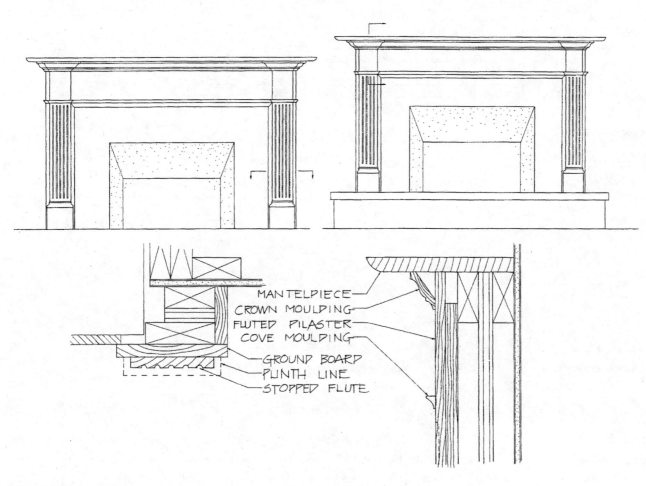

**Figure 13-4**
*An elevated hearth may require a reduced scale of woodwork (right).*

The height of the fireplace hearth affects woodwork design as a whole, as an elevated hearth will, by necessity, elevate associated woodwork by a like amount. Elevated hearths are often incorporated into fireplace designs to provide low seating and to make the firebox more accessible and easier to tend. An increase in overall woodwork height may require a slight decrease in moulding scale or proportion (see Figure 13-4), in order to maintain a correct appearance on the wall and in the space.

Likewise, the size of the firebox opening may affect woodwork scale or proportion to a certain extent. An oversized opening tends to enlarge or spread out the fireplace woodwork, possibly creating a sparse appearance. Conversely, a minute firebox is usually not the place to mount an overwhelming piece of classical finery, as the firebox might appear stifled or lost.

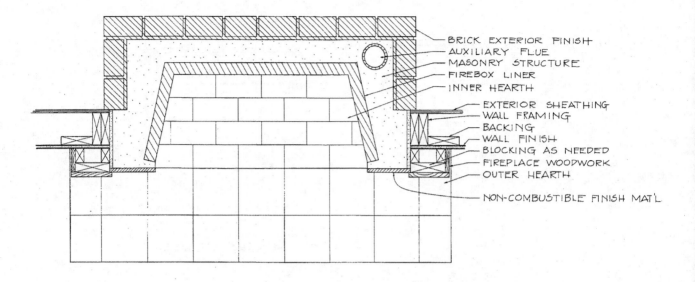

**Figure 13-5**
*Plan view cross-section of a fireplace, showing woodwork and blocking.*

Finally, the practical matter of providing for efficient and secure attachment of woodwork components must be confronted at the rough phase of fireplace construction. Often this step amounts to simply adding additional framing pieces around the fireplace perimeter, or it might involve the embedding of wooden nailers into the brick or CMU surface as the fireplace is constructed. Good planning here pays off with a smooth and positive installation of woodwork at the trimming-out stage.

## Woodwork Designs

Figures 13-5 through 13-9 illustrate a series of fireplace woodwork designs and details, progressing from a minimalist cap moulding to complex designs that draw directly on the design of the classical orders (as applied to openings).

# Fireplace Woodwork

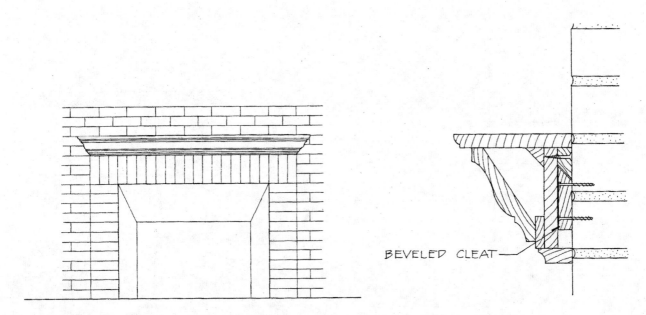

**Figure 13-6**
*An independent mantel mounted directly onto masonry.*

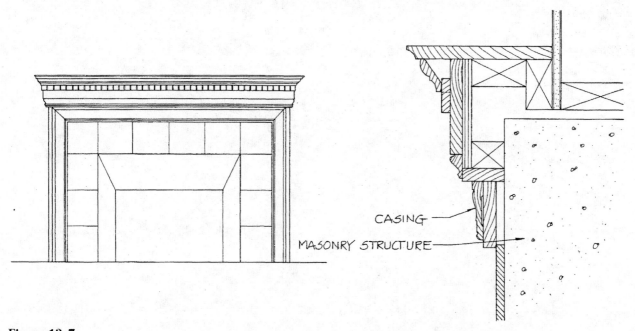

**Figure 13-7**
*A cased opening with mantel above. The casing should, ideally, echo adjacent window or door trim.*

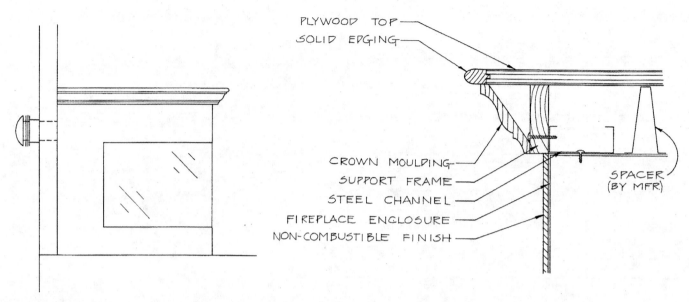

**Figure 13-8**
*A bar top on a direct-vent peninsula fireplace.*

**Figure 13-9**
*A wood mantel on masonry corbels.*

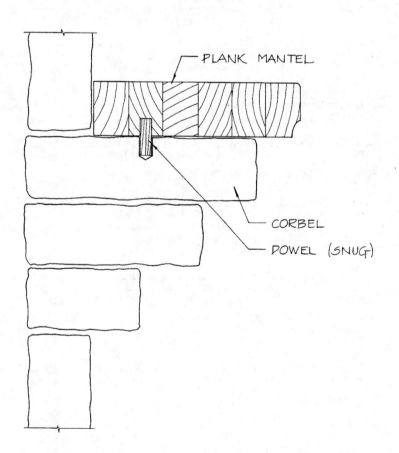

CHAPTER 14 | # Cabinetry and Shelving

Permanent interior architectural elements designed for storage include cabinetry (formal) and shelving (informal); today these are generally one of two basic types: modular or site-built. Modular components, whether mass-produced in a factory or from a limited run in a small local custom shop, are intended for rapid installation and relatively easy alteration. Each separate preassembled piece (module) is fixed in place, to create a whole from several component parts. By its nature, manufactured modular cabinetry and shelving is usually available within a limited range of standardized dimensions.

Site-built cabinetry and shelving is distinguished from modular components by its custom-designed, pieced construction. It is assembled and installed to fit specific size, space, and construction restrictions for a particular application, and it is constructed in place, en masse, as an integrated whole.

Regardless of the type of construction, the design of storage woodwork should be approached with an understanding of basic practical and ergonomic conventions as well as woodworking materials and methodologies. This chapter presents a selection of construction and specification details for several types of storage woodwork, including kitchen cabinetry, general casework, and clothes closets.

## Conventions

Storage woodwork is interactive woodwork—it is touched, manipulated, filled, and emptied by people. Like doors, stairs, and other functional woodwork, storage space must be designed with its ergonomic ramifications in mind. Building codes tend to recognize this and have adopted standards for dimensions that strive to provide safety and convenience for average users of certain furnishings and work surfaces.

In addition, standard dimensions allow for the efficient manufacture of accessory elements. Sinks, for example, are produced in a limited range of widths intended for installation in kitchen and bath countertops of standard sizes. Built-in appliances like ranges and dishwashers are commonly available in only one height, to fit inside standard-sized cabinetry.

Designs that call for dimensions outside these standards should be carefully considered in light of budget constraints, subsequent use by later occupants, and the availability of accessories that mesh well with odd layouts. Handicapped individuals often need very low work surfaces that can be conveniently manipulated from the confines of a wheelchair. Very tall people, or those with chronic back problems, can benefit by the use of higher-than-average work surfaces.

Figure 14-1 illustrates accepted standard dimensions for various cabinetry and shelving arrangements. Note that *height* standards are based most strongly on ergonomic design criteria—heights at which an average human can easily reach and manipulate surfaces and spaces for particular uses. Depth standards, while closely tied to height criteria and certainly possessing some ergonomic ramifications, tend to be based more on the economics of production: achieving maximum yield from standard sheet goods. Depths of 12, 16, and 24 inches divide nicely into common 48 × 96 inch sheet goods.

## Modular Components

In most cases, the design of cabinetry assemblies and layouts that will utilize modular components is fairly straightforward. Once the basic space and use requirements have been established, one can, with the manufacturer's catalog in hand, insert appropriate cabinets that fit the space and meet the needs of the user. In the case of kitchen cabinetry, a good basic knowledge of accepted

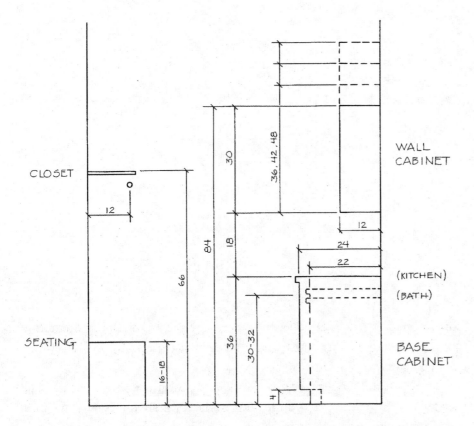

**Figure 14-1**
*Conventional dimensions for storage woodwork.*

kitchen design principles is necessary to ensure an appropriate configuration. Proper positioning of appliances and fixtures is crucial to a successful kitchen design, and the adept positioning of standard-sized components in a nonstandard space is the hallmark of a good kitchen designer.

Manufactured cabinetry, always modular by its nature, is commonly marketed in the U.S. as either "stock" (off the shelf), "custom" (made to order), or "semicustom" (a combination of the two). (Knock-down cabinetry, which is delivered unassembled, is not discussed here.) These subdivisions tend to be more ploy than substance, however: the vast majority of kitchen and bath layouts can be readily served with stock cabinetry, which is generally available in a wide range of sizes in three-inch increments. The advantages of the custom lines come into play when special sizes and internal fittings are required; even so, such specialty work is often accomplished more neatly and at less cost by site- or shop-building the cabinetry instead.

Until recently, nearly all cabinetry in this country was built of *face-frame* construction, detailed in Figure 14-2. Each cabinet is constructed of relatively thin panels and is faced with a frame of solid stock that provides strength, finishes the panel edges, and allows for the mounting of hardware. Unfortunately this method of construction, based on some principles of furniture construction

**Figure 14-2**
*Face-frame cabinet construction.*

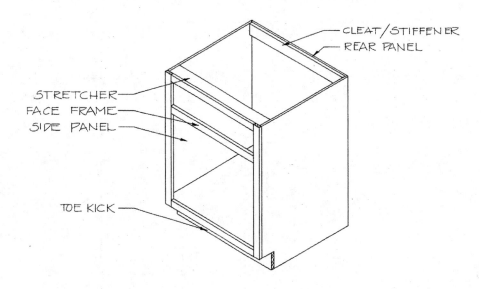

and the tradition of site-built work, is not the most economical method of mass-producing cabinets, in light of modern materials and manufacturing technologies.

So-called European, or frameless, cabinet construction—really an older technique known as carcass construction to the furniture trade—utilizes sheet goods, namely industrial particleboard, in a highly efficient manner to produce minimalist boxes to which doors and drawers are fitted using specially-designed hardware. Construction details of frameless cabinetry are illustrated in Figure 14-3.

**Figure 14-3**
*Frameless cabinet construction.*

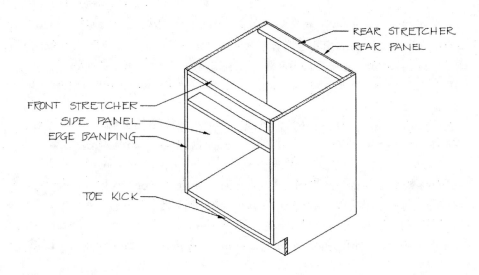

# Cabinetry and Shelving

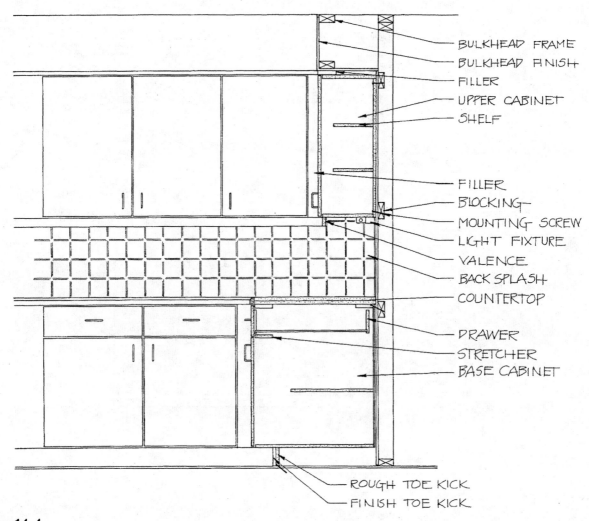

**Figure 14-4**
*Installed modular cabinetry: a typical kitchen elevation and cross-section with flush soffit (bulkhead) above.*

Frameless cabinetry, unencumbered by the contrasting vertical and horizontal bands of the face frame, is capable of providing very clean lines; depending on the door style and edge conditions, is adaptable to many contemporary and traditional schemes. And since it is readily and inexpensively manufactured, by mechanized processes and largely of by-product material, European cabinetry has secured a substantial segment of the cabinet market.

Installation details of modular cabinetry are illustrated in Figures 14-4 and 14-5. Each cabinet is secured in place by screwing it into the wall frame through the wall finish material and attaching it to adjacent cabinets with screws or special studs. Some cabinet manufacturers offer special tracks for rapid and secure mounting of wall cabinets. Base cabinets are capable of

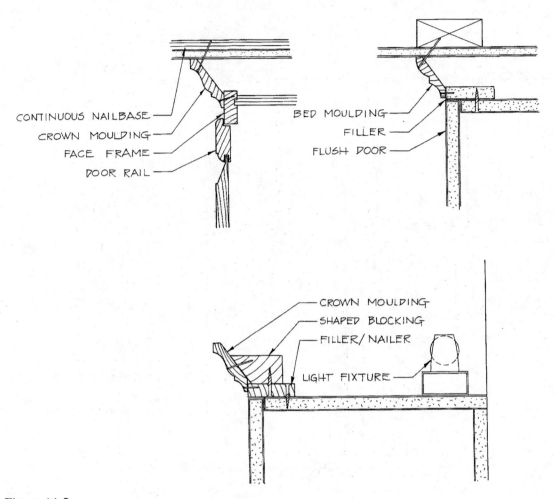

**Figure 14-5**
*Terminations at top of modular wall cabinetry.*

holding large vertical loads, since their strength is bolstered by the floor. Wall cabinets, however, must be well attached with appropriate load-bearing fasteners to prevent heavy items (like dishes) from collapsing them.

The periphery of an installed cabinet, like all edges, merits special attention. Various fillers, trims, and mouldings are often required to smooth and complete a modular cabinetry installation. A run of stock cabinets between two walls may come up a bit short, and the resultant gap must be covered with a filler strip. The uppermost edge of wall cabinetry must be well planned, especially if it's complicated by the presence of a drywall soffit or a crown moulding. Framing and wall finishes are best undertaken with the cabinetry plan in hand, to ensure a trouble-free and natural-looking cabinet installation later. The lowermost edge of base cabinetry deserves special attention, too. An adequately proportioned toe space and a durable kick plate are necessary whenever cabinetry will be used to support a work surface, such as a kitchen countertop. The specifics of the finish flooring and the timing of its installation are very important, and should be known before the cabinetry is designed or laid out. In most cases it is the most prudent practice to install modular cabinetry *over* finished flooring or underlayment (or appropriate blocking) of any substantial thickness, both to avoid diminishing the toe space and to ensure that undercounter appliances will fit.

## Site-Built Cabinetry

Sometimes the use of easily installed, readily available modular cabinetry is supplanted by special needs: an unusual space, highly specific construction requirements, handicapped access, or the desire for characteristic or unusual work. Site-built storage can be readily designed to fit such specific requirements. In addition, prevailing labor rates may make site-built installations more cost-effective than custom-made modular counterparts.

Since it is the means of construction that differentiates site-built work from modular installations, the design principles governing site-built work are essentially the same as for modular design: ergonomic considerations are the first priority, kitchen design principles are applied equally, and edge conditions, though less troublesome, must, as always, receive special consideration.

Installation details of some site-built cabinetry are illustrated in Figures 14-6 and 14-7. Note that since individual modules are not necessary, a material savings can be realized by using shared side panels for adjoining cabinets. And although they are desirable, separate panels for the backs of cabinets are not always needed, as the existing wall finish may serve that purpose adequately.

## Closets

Residential closets, generally small enclosed rooms for clothes storage, are subject to similar ergonomic and practical design criteria as for storage cabinetry and work surfaces. Several options are available to fit out closets for efficient storage. An entire closet accessories industry exists, providing "systems" of coated-wire shelving and storage, or laminated panel products and

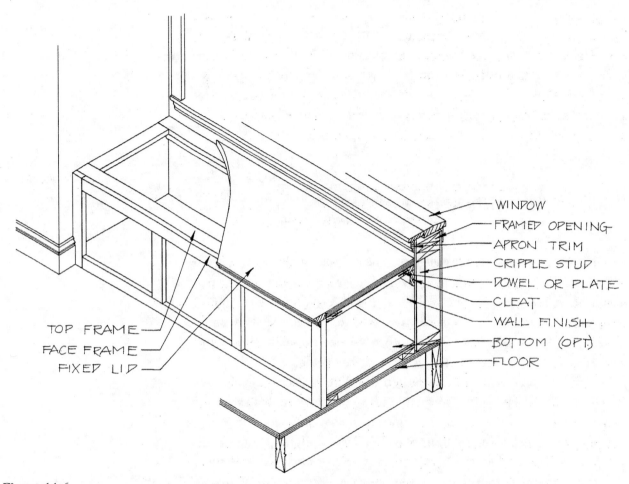

**Figure 14-6**
*Installed site-built cabinetry: a window seat (shown without doors).*

related fittings. This section illustrates some details of site-built closet storage solutions and design standards.

In order to accommodate typical clothing suspended on wooden or wire coat hangars, a closet should be at least twenty-two to twenty-four inches deep; closet rods should be placed at the midpoint of this depth. Shelving positioned above the hangar rod should not protrude so much as to block overhead lighting or interfere with the manipulation of the hangers. A typical rod-and-shelf closet is illustrated in Figure 14-8.

A two-tiered closet configuration is detailed in Figure 14-9. Note that the upper rod needs to be positioned higher than the typical sixty-six inches, in order to accommodate two runs of clothing. The shelf must likewise be higher, and it thus will likely see less frequent use and should be used for storage of infrequently used items.

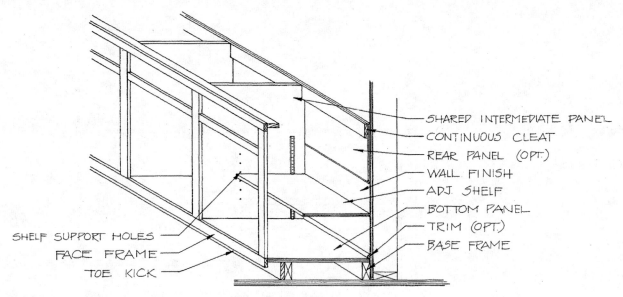

**Figure 14-7**
*Installed site-built cabinetry: a run of kitchen base cabinets. Note the shared panels and the option of omitting a separate back panel.*

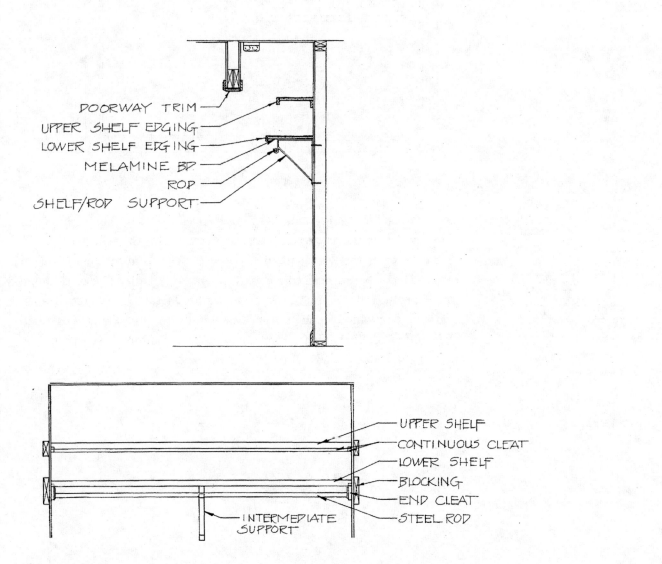

**Figure 14-8**
*Typical closet construction in cross-section (above) and elevation.*

**Figure 14-9**
*Two-tier closet construction.*

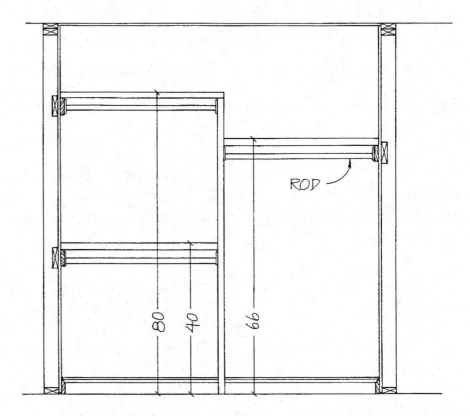

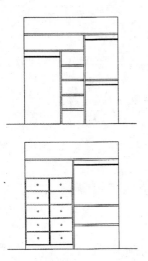

**Figure 14-10**
*Additional closet layouts.*

Beyond the basic combinations of shelves and rods, closets can be fitted with various drawers, receptacles, and hardware for the storage of specific items such as shoes, lingerie, ties, and accessories and to provide for an organized, compartmentalized approach to the storage of everyday clothing. The possibilities are limitless; two examples are illustrated in Figure 14-10. It is important to note, though, that closet woodwork is generally considered to be informal; it is hidden behind a closed door and therefore need not be built to the same standards as exposed cabinetry.

CHAPTER 15 | # Stairs and Balustrades

Like doors, stairs are essentially utilitarian—a series of ascending steps that allows for easy passage from one level to another, normally accompanied by a handrail or handhold for safety. But however functional its purpose, stair design, like door design, can provide an unlimited variety of stylistic decorative interpretations, often quite focused and traditionally ornate.

This chapter presents a series of construction details for some primary types of staircases and balustrades (rail systems), as well as pertinent dimensioning and proportioning specifics.

## Stair Construction

Traditional *housed* stair construction—named for the joints where the risers and treads are recessed into a groove, or "housing," in the stringer—is illustrated in Figure 15-1. Structurally this stair assembly is a marvel of interdependent elements, each acting in its own way to achieve the load-bearing potential of the whole. The stringers act as long beams spanning the distance between floors. Across these beams are placed the risers, which assume the role of short joists or purlins spanning the distance between the stringers. Finally, treads laid atop the risers and into the stringers form the working platform of the staircase. Strategically positioned wedges, glue blocks, and mouldings complete and tighten the assembly, which should be designed to provide a solid means of ascent and descent for the life of the building.

**Figure 15-1**
*Housed stair construction.*

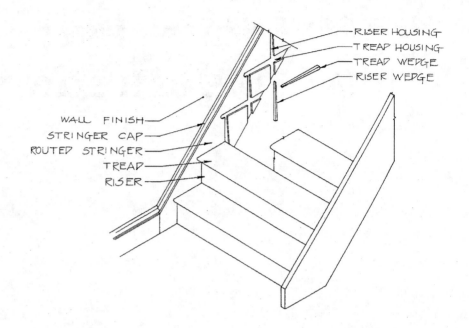

From this basic interdependent structural design several variations stem. A common permutation is the open stringer, which may occur on one side or both sides of a run of stairs, for part or all of its length. It is illustrated in Figure 15-2. It differs from the closed, housed stringer of Figure 15-1 in that the open stringer is notched to receive each step (tread and riser), exposing the familiar saw-tooth profile of stairs in elevation view. Once a stringer is notched in this manner, allowances must be made for the loss of its span capability, due to effective decreased depth of the notched member (see Figure 15-3). The notched stringer may be doubled, thickened, or deepened; or it may be strengthened by constructing a wall beneath it. This lat-

**Figure 15-2**
*Open stringer construction.*

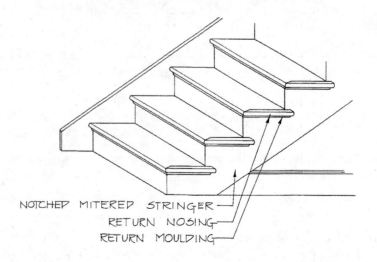

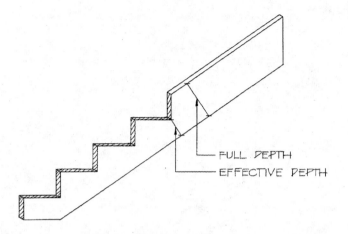

**Figure 15-3**
*Notching a stringer (or beam, joist, and so on) decreases its effective depth, weakening it.*

ter detail is very common, as the space beneath stairs is rarely left exposed; rather it usually encloses another, lower, stair or a closet.

An economical variation on the original utilizes rough "horses," or stringers of dimensional framing lumber, to provide the primary longitudinal support for the stair (see Figure 15-4). The risers and treads are simply applied to the notches of the rough stringer, eliminating the more labor- and skill-intensive chore of housing. This construction is often used by production builders and the stairs later covered with carpet. A tight tread joint at the finished stringer is not necessary; if its joints are well fastened and glued and of good material, this stair is durable and adequate. It should be noted that rough notched stringers are recommended for the center or intermediate

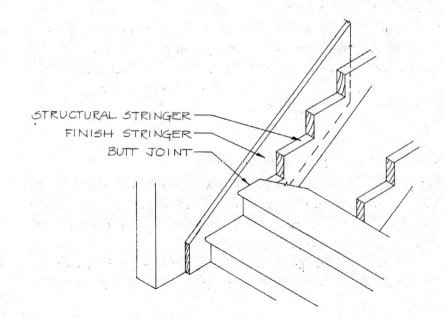

**Figure 15-4**
*A semihoused stair uses notched stringers of framing lumber for primary support.*

**Figure 15-5**
*An open staircase of large-section timbers.*

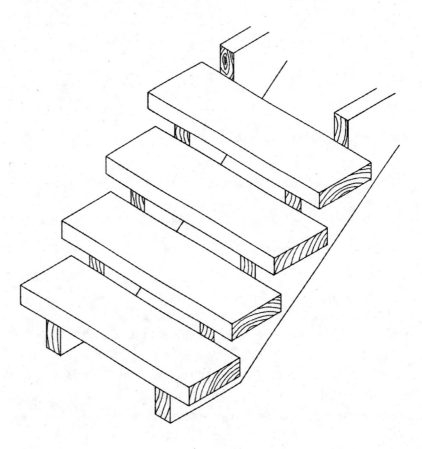

stringers on most stairs (housed or otherwise) greater than about forty inches wide, as the span capability of a (nominal) one-inch riser begins to be compromised beyond that span.

Another category of stair construction is the open case, in which the risers are eliminated, leaving only treads spanning the stringers. This construction is common in some contemporary and commercial designs, and if well done it can add a feeling of openness and airiness to a space. A typical detail is illustrated in Figure 15-5.

## Stair Configuration and Layout

The simplest, most economical, and most user-friendly stair configuration is the straight run, shown, with other configurations, in Figure 15-6. The straight-run stair provides safe, uninterrupted ascent and descent, and it requires relatively simple construction techniques.

If the building design is such that a straight run will not fit, then another configuration is necessary. The split flight, turned either at 90

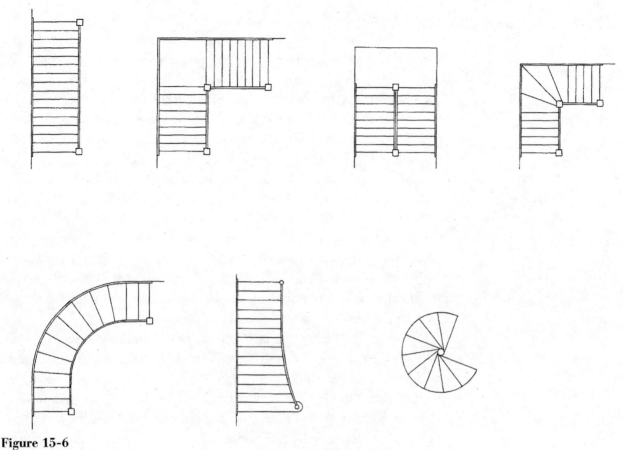

**Figure 15-6**
*Typical stair configurations. Top: a straight run; a 90° turn; a 180° turn; a 90° turn with winder treads. Bottom: a curved stair; a flared stair; a circular stair.*

degrees or 180 degrees, may allow a stair to occupy building space more favorably. The necessity for a small, intermediate floor along the flight (a landing) tends to make these stairs a bit more complex and costly to construct and somewhat more difficult to use.

If space is extremely tight, then a circular staircase (often called a spiral staircase, which is a misnomer as it is really helical) may be constructed. Circular stairs, however, are very difficult to use, and they should not be designed into spaces as the primary means of ascent and descent unless the space will be used on a limited basis.

If aesthetics precede budget on the priority list, then the addition of decorative flaring curves is an excellent means of providing beauty and elegance to many stair layouts.

## Conventions

Most architectural woodwork is designed and constructed around the human scale. Some woodwork, namely doors, cabinets, and (especially) stairs, really only work well when the ergonomic consequences of their use are fully considered.

Experience has shown that a comfortable ascent and descent along a run of stairs will occur only within a very narrow range of dimensions for the risers and treads. These dimensions are called the *rise* and the *run*. From a practical standpoint, a steep stair (large rise, small run) is difficult to climb and unsafe to descend. A shallow stair, while easy to ascend, consumes a lot of space and causes too much forward motion relative to vertical motion, making the ascent and descent seem inefficient. Several rules of thumb exist for designing the rise and run of stairs, and building codes normally incorporate some of these into their standards. Optimal stair design dimensions are subject to debate. The codes are always evolving, and local requirements should always be checked when designing any nontypical stair. Table 15-1 lists the common ranges for rise and run configurations.

Generally the rise of a step (the "unit rise") should not exceed 8 inches; a greater rise tends to be difficult for many to climb. A rise in the range of 7 to 7½ inches works well. Although a unit run of a step should be greater than 9 inches, a minimum of 10 inches is even better, and for outdoor use, a run of 12 or 13 inches provides for safer foul weather navigation. Figure 15-7 illustrates the differences in several stair layouts—note that tread width is typically different from the dimension of the run, as the tread typically overhangs the riser a certain amount as a nosing. The nosing provides for a safer, more natural climbing and descending motion.

### Table 15-1 Stair Conventions

| Factor | Ideal Value | Minimum Value | Maximum Value |
| --- | --- | --- | --- |
| Unit rise | 7" (interior) | 4" | 8.25" |
|  | 6" (exterior) |  |  |
| Unit run | 11" (interior) | 9" | 14"+ |
|  | 13" (exterior) |  |  |
| Slope angle | 32° (interior) | 20° | 42° |
|  | 24° (exterior) |  |  |

# Stairs and Balustrades

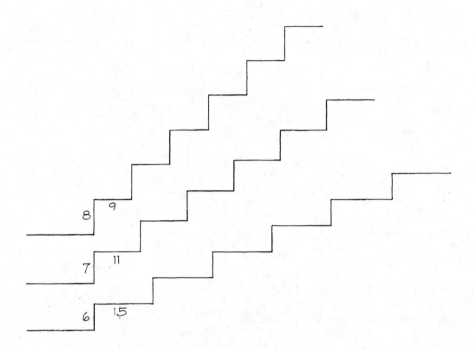

**Figure 15-7**
*The effect of rise and run variation.*

In general, frequently used stairs should be designed with a good balance of comfort and efficiency. Infrequently used stairs (perhaps for a basement or attic) may, as a means to conserve space and material, be designed toward the steep end of the dimensional spectrum; likewise, outdoor stairs should be designed toward the shallow end for reasons of safety.

## Rails and Balustrades

For safety, virtually all stairs of more than three risers should have a handrail. Closed stairs, bound by walls on both sides, need at least one rail, usually supported by brackets or a cleat at a height prescribed by code (see Figure 15-8).

Open stairs call for the construction of a more elaborate rail system, as the rail must be well supported without benefit of a wall, and the space beneath the rail must be blocked to prevent falls. Figure 15-9 illustrates the components of a typical wood balustrade.

With the exception of a newel-less balustrade, such as in Figure 15-10, the main supporting members of a balustrade are its newel posts, or simply newels. Typically installed at either end of a run of rail and at each

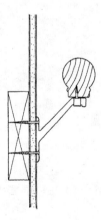

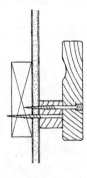

**Figure 15-8**
*Applied handrails.*

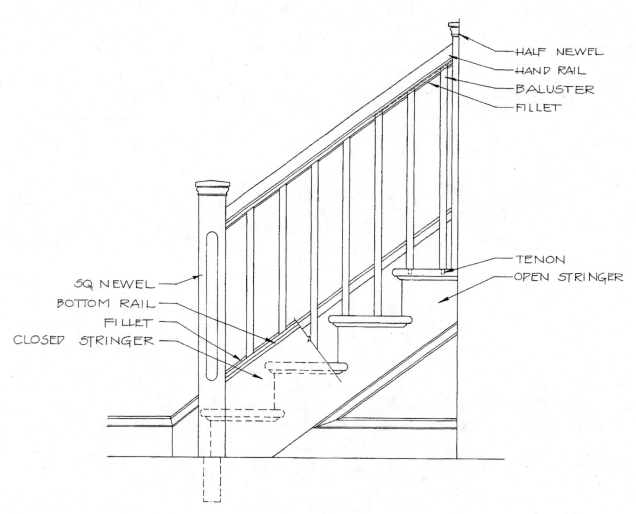

**Figure 15-9**
*Balustrade components, showing a closed stringer (left) and open stringer.*

landing or corner, newels resist vertical as well as lateral force against the rail. Spaced balusters, also termed spindles (if turned), provide secondary support to vertical rail loads as well as a blockade to possible falls.

A variety of stock manufactured balustrades are readily available in a limited range of traditional and contemporary styles. But since stairs and rail systems provide an inviting opportunity for creative design, these common parts are often forsaken for a bit of creative, often signature, woodwork on the main stair of a building.

A range of suitable balustrades is illustrated in Figure 15-11.

## Stairs and Balustrades

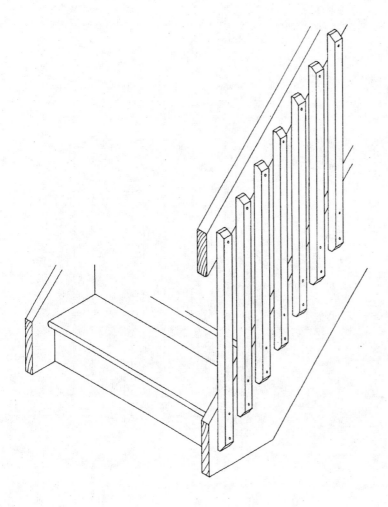

**Figure 15-10**
*A contemporary newel-less balustrade. Lateral strength is developed at screwed connections and the repeating bearing surfaces of each baluster on the face of the stringer.*

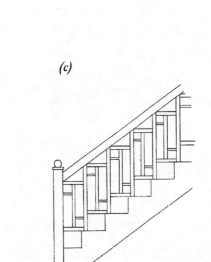

**Figure 15-11**
*Additional balustrades. (a) Steel with wood.
(b) Turned balusters and newels with continuous rail.
(c) Custom grid-work. (d) Neotraditional with double stringer and bottom rail.*

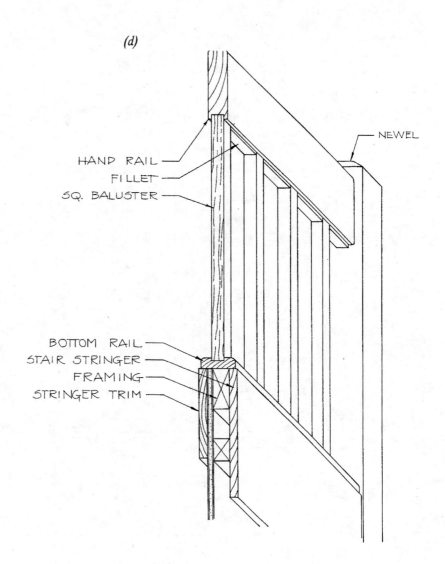

**Figure 15-11**
*(continued)*

# Section Four
## Special Topics

# CHAPTER 16 | Curves

Throughout history, curves have been used as architectural elements in one form or another to lend beauty, visual impact, or function to building design. Curved forms include arches, such as for passageways or portals; plan elements, such as curved hallways and walls; and miscellaneous elements, such as barrel vaults or elliptical stairs. If well used, curves can add spectacular beauty, succinct elegance, and the ultimate simplicity of form to any building design, no matter the scale.

Naturally, there is a price to pay for such a versatile design tool—the price of production. The timeless builder's adage "curves cost" reflects the additional time, skill, and tooling necessary to make the significant leap from flat, square, and straight to curved, arched, and round. However simple they may appear as finished products, curved elements will, without exception, cost more to produce than their square or angular counterparts. The additional expense of curved work applies to all building trades; fortunately, woodwork is quite suitable for the presentation of many curved forms due to its working versatility.

Curved woodwork can be laid out in any of the conic sections: circular, parabolic, hyperbolic, and elliptical. Circular and elliptical forms are the most commonly used curves in modern wood design—the former because they are easily defined and replicated, and the latter for their graceful elegance.

This chapter describes and illustrates some methods of dealing with the presentation and production of curved details constructed of wood or wood substitutes, to allow the designer or builder to make informed, practical decisions pertaining to the specification and execution of curved work.

# CHAPTER 16

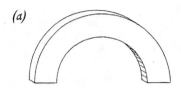

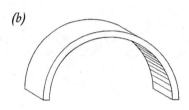

**Figure 16-1**
*The two basic categories of woodwork curves. (a) A long-dimension curve. (b) A short-dimension, or natural curve.*

## Types of Curvature

The techniques used to cause wood to assume a curved shape can be split into two basic categories. The direction of curvature relative to the cross-section of the curved member defines each category. Assuming the wood has a rectangular cross-section (most boards do) then the curvature of a curved board is a deformation of either the *long* cross-sectional dimension or the *short* cross-sectional dimension. See figure 16–1 for an illustration of these two categories.

These two basic types of curve often requiere significantly different techniques to produce. Long-dimension curves are somewhat unnatural. They represent curvature that is essentially impossible in a solid piece of lumber; therefore their execution requieres a relatively complex construction or fabrication. Short-dimension curves, on the other hand, represent the natural directin of curvature a board will take with applied bending force. Short-dimension curves follow a plane of weakness; their execution is often a simple matter of assistance or persuasion.

### Long-Dimension Curves

This section considers curves along the long dimension, in a series of three examples. The first example is a flat exterior casing to trim a semicircular window. The second is a curved band to end a wood floor. The third is the top of a curved deck rail. Alternate techniques for constructing each of these examples are discussed.

Figure 16-2 illustrates an elevation view of a wood window trim (casing) surrounding a semicircular top window.

Three methods, all buildable on-site by siding carpenters, are recommended for this application. The first, and best, is to build a multipiece casing formed of three or four short, wide boards, joined well with reinforced butt joints and carefully cut to the appropriate radius. Figure 16-3 illustrates the essential features of this method. Dowels or plates (biscuits) may be used for the joint reinforcement, well glued with waterproof adhesive. If the work is to be placed high up and not subject to close scrutiny, the curve may be sawn. If it will appear at eye level then an extremely accurate curve may be simultaneously struck and cut with a router on a trammel.

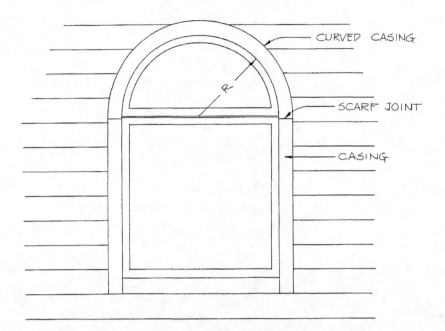

**Figure 16-2**
*A curved exterior window casing.*

A simpler method, likely to be less durable, is to cut each segment of the curve separately, and simply nail the segments into place. The plain joints will likely open, however, and may allow detrimental water infiltration.

The third method, quite tempting for its simplicity, is to cut the curve, in one piece, out of a sheet of exterior plywood of suitable thickness. This method should be reserved for opaque (paint) finishes only, for two reasons: the

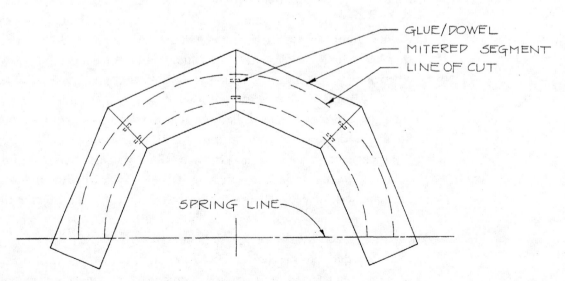

**Figure 16-3**
*The construction of a multipiece arch with reinforced joinery.*

**Figure 16-4**
*Double-sealed flashing.*

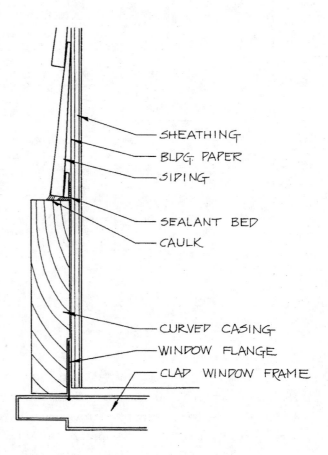

exposed edges of the plywood need an extremely durable and weather-resistant finish in order to effectively turn away water, and the unsightly unidirectional grain in the multi-ply structure of plywood is masked with an opaque finish. Caution should always accompany the exterior use of plywood—extreme care must be used to completely seal its edges and faces (all around) from water; otherwise it *will* eventually degrade or delaminate the material over time.

Typically the head casing of a window is flashed to turn water away. On a flat casing, a simple Z flashing (Chapter 3) works very well. Such a flashing is difficult to fabricate into a curved form, however. Soft copper sheet, cut and soldered, is an effective, albeit time-consuming, option. A more economical technique, effective if properly executed, is the double-caulk method. Immediately prior to the installation of siding atop the curved casing, a generous bead of high-durability caulk is placed at the joint where the casing meets the sheathing (see Figure 16-4). The siding is then installed and embedded into the fresh caulk. When the entire window is finished, the siding is then caulked to the curved casing again, creating a double seal.

Figure 16-5 illustrates the plan view for a terminal band trimming the end of a curving wood floor.

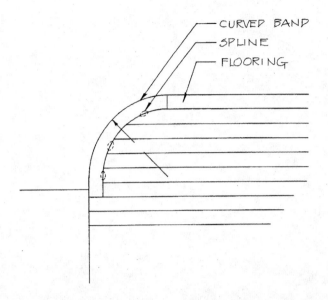

**Figure 16-5**
*A curving terminal band on a wood floor.*

Two techniques are recommended for this application. Again, the first and simplest is to build up the curved band with short, wide mitered boards and reinforced joints. For a tight, well-fitting joint between the floorboards and the trim band, a router, on a trammel or against a template, should be used to cut the floorboards and the curved trim pieces. (Any unguided technique will leave unsightly gaps.) Once cut, the curved trim pieces should be joined to the floorboards with a spline or plate joint (see Figure 16-5), glued to the curved band only so the flooring will be free to contract and expand with seasonal changes in humidity. The curved pieces may be nailed or screwed to the substrate.

A second technique, which is a bit more involved than the first, is to use bent laminae (layers). Thin strips of wood are simultaneously bent and glued in such a way that the finished piece permanently assumes the curved shape desired (see Figure 16-6). The advantages of this technique are that the curve is precisely formed and does not need to be cut, and the grain is smooth and continuous with the curve, so the wood appears to be naturally curving. The straight, stop-and-start grain of segmented work is eliminated.

If conditions allow, the curved end of the flooring provides a suitable form to bend the laminae against. Wedges fastened to the subfloor can act as clamping members to hold the assembly as the glue dries (see Figure 16-6b). Otherwise, the laminations can be glued in a form, with clamps—a standard mill-shop technique (see Figure 16-6a).

Figure 16-7 illustrates a detail and plan view for a curving exterior deck rail. The subrail is to be constructed using the bent lamination technique described above. Three techniques are recommended for forming the wide top rail. The first is to, again, use bent laminae. This technique will provide a

# CHAPTER 16

**Figure 16-6**
*Bent lamination.*
*(a) Using a shaped form.*
*(b) Using floor as bending form.*

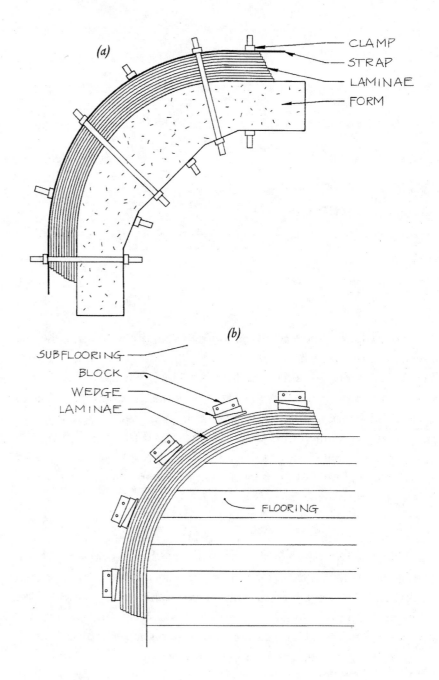

finished piece of exceptional strength, continuous grain flow, and good appearance. It is the costliest choice, though, due to the added expense of constructing clamping forms and the large amounts of exterior glue needed.

A second technique, termed stack lamination, will provide the strength this piece requires, but it is somewhat less expensive. Figure 16-8 illustrates the basic features of this method. Two or more layers of wide, mitered stock are stacked together, joints well offset, to bring the piece to the final desired thickness. In this case, three ½ inch layers are combined to form

# Curves

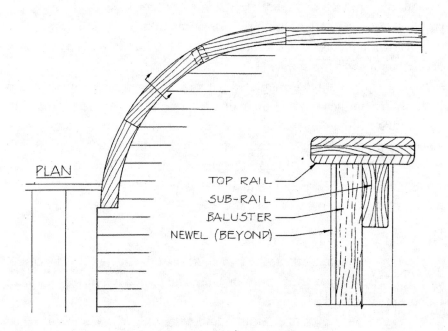

**Figure 16-7**
*A curving deck rail in plan and cross-section views.*

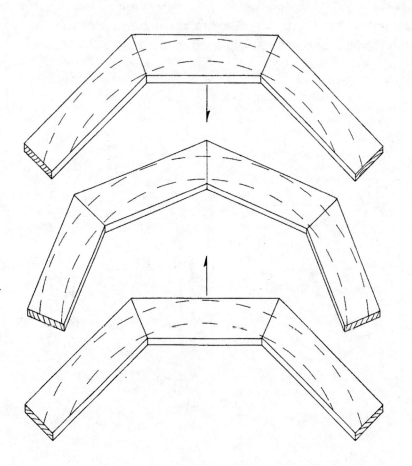

**Figure 16-8**
*Stack lamination.*

# CHAPTER 16

a nominal 2 × 6 inch rail. Once clamped and set, the stack can be cut to the required shape, smoothed, and installed; no moulding form is necessary. The only real shortcoming to this technique is the stop-and-start grain flow, most visible on the rail's top surface.

A third technique is suggested purely as a cost-controlling method, as it results in a finished piece of limited independent strength. This is the technique using multiple mitered pieces, already described in previous examples, where short, wide stock is joined with dowels, splines, or plates and well glued. This method depends on the subrail and newels to lend adequate strength to the rail—each joint should fall over solid support.

## Short-Dimension Curves

Three examples illustrate the techniques used to achieve short-dimension curves. The first is an arched doorway jamb. The second is a curved two-piece baseboard. The third is a curved wainscot panel.

Figure 16-9 presents elevation and cross-section views of an arched doorway. The casing is produced by one of the long-dimension techniques described earlier. The jamb may be formed by bent lamination or by two alternate techniques: kerf bending or steam bending.

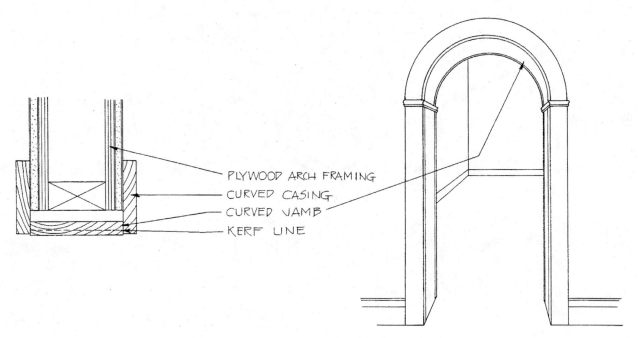

**Figure 16-9**
*An arched doorway.*

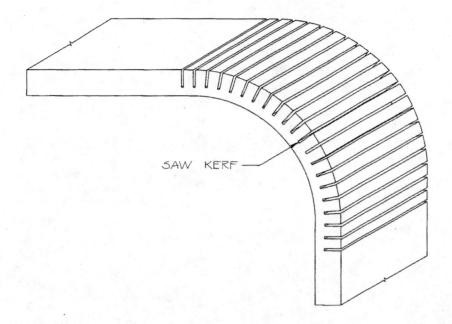

**Figure 16-10**
*Kerf-bending.*

Kerf-bending is an old and proven method; a series of partial saw cuts, called kerfs (see Figure 16-10), across the grain will weaken the piece sufficiently to allow bending it, either toward or away from the kerfs. Kerf-bending is well suited to a piece that will have just one of its faces exposed to view, as our jamb will. Once the piece is installed, the saw kerfs are hidden by applied trim. A kerf-bent piece is also severely weakened by the process, and is ideally suited for designs that offer the bent member secondary support or backing.

Other bending methods retain the full strength and integral appearance of solid stock. One such method is steam-bending, which entails just what its name implies—subjecting the wood stock to severe temperature and moisture stress as a temporary weakening measure. Once wood is so weakened, it may be bent to a form or into a freeform shape, and it will cool and hold its new configuration with excellent integrity.

Typically, the stock to be bent is placed in a chamber (a long pipe or box), and steam is pumped in for a specified time period, which varies according to species, thickness, and the severity of the intended bend. Normally, steam-bent wood is overbent by a certain percentage, as it will tend to spring back toward its flat state when the bending pressure is released.

A third method, already presented as a long-dimension bending process, is the more controllable and predictable bent-lamination technique. Thin layers of stock are glued and clamped to a form; once set and cleaned

**Figure 16-11**
*A curving baseboard in plan and cross-section views.*

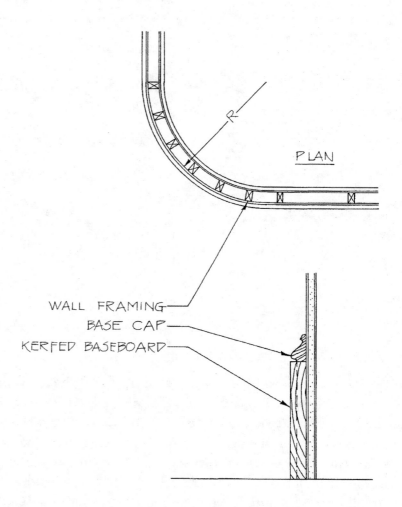

they appear virtually as one. The laminae do have a tendency to spring back to their original form, however, and a gradual process called glueline creep can also cause a loss of curvature accuracy. These problems can be of major consequence in unsupported applications such as furniture, but they should not be a concern in architectural woodwork, which is normally attached to a substructure.

Figure 16-11 provides plan and cross-section views of a curving baseboard composed of a flat main section topped by a traditionally moulded base cap. The flat section can be kerf-bent, as it will have only one finished side. The base cap, however, presents a more complex problem, as its moulded shape prevents it from bending smoothly without excessive deformation in the vertical plane. Moulded shapes can be steam-bent, although the tendency for the grain to raise during steaming may require extensive sanding, which could remove a part of the moulded profile. A better technique, requiring considerable skill and attention to detail, is to use a modified bent lamina-

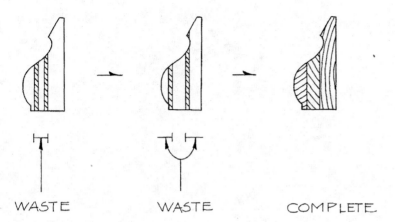

**Figure 16-12**
*Bent lamination of moulding.*

tion, ripping the moulding into thin layers and gluing it back together in a curved orientation. The ripping process, by its nature, will remove a saw blade–width kerf from the profile, lost into sawdust. This technique thus requires two lengths of moulding, each ripped carefully so as to end up with adequate layers of the complete profile (see Figure 16-12). The layered moulding can be glued along a form, or possibly right in place. Care must be used to preserve the correct orientation of each layer, so the profile is true when the layering is complete.

Figure 16-13 provides plan and cross-section views of a curving wainscot, composed of raised panels mounted behind curved stiles and rails. The best method to produce curves along wide stock is to use a staving technique, named after the narrow stock forming the walls of wooden barrels or casks. Beveled strips are glued up to form a rough version of the required curve; once these are set, the final shape is attained using curved planes by hand or with guided routing.

# CHAPTER 16

**Figure 16-13**
*A curving wainscot in plan and cross-section views.*

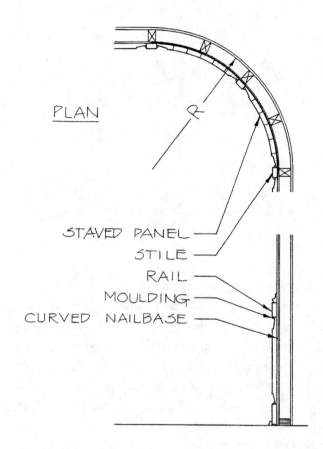

# CHAPTER 17 | Exterior Protection

*A little caulk, a little paint, makes a carpenter what he ain't.*
—M. Vargo

Exterior architectural woodwork consists, in large part, of a variety of cleats, claddings, and constructions installed to provide basic protection from the harmful effects of the elements for vulnerable structural components and interior finishes. In turn, exterior woodwork needs its own systems of protection so that it may provide prolonged service and remain attractive for the life of the building. Most wood, once removed from the relatively safe confines of a living tree, is highly susceptible to attack and degradation. If used as an exterior building material, wood must therefore be safeguarded against premature failure.

Rain, snow, extremes of temperature, and particularly the powerful rays of the sun all conspire to degrade and damage exterior wood. This chapter describes two categories of exterior protection which can decrease these damaging elements: design protection (passive) and applied protection (active). Often utilized in combination, both classes of protection are important to many types of exterior woodwork.

## Design Protection

Whenever possible within the confines of the overall building design, provisions should be made to provide built-in protection of exterior woodwork and the structure beneath. A variety of appropriate design details, both large and seemingly insignificant, can decrease the degrading effects of one or more harmful elements. The most significant of these are 1) ultraviolet shading, 2) pitch for runoff, 3) water-eliminating joints, 4) flashings, 5) resistant materials, and 6) effective landscaping.

## Ultraviolet Shading

Sunlight, particularly if it is direct and prolonged, bombards wood fibers and the coatings we apply to them with enough ultraviolet and other radiant energy to cause detrimental breakdown of the relatively stable chemical composition of exterior woodwork and finishes. Oxidation (with loss of wood mass) is accelerated under such conditions, finish coatings are stressed enough to fail prematurely, and wood movement is exaggerated by the presence of excessive heat. To varying degrees, each of these detrimental effects conspires to cause failure of exterior wood components, resulting in the need for more frequent maintenance and repair. Thus the elimination of direct exposure to radiant energy is an effective means of prolonging the life of exterior wood.

For example, the extension of a roof, in the form of an overhanging cornice, provides a degree of protection for the wall woodwork beneath it, from both sunlight and precipitation. This basic form of shading is not perfect, of course, and its effectiveness depends on many factors, including the size of the overhang and climatic variables like wind direction and sun angle. As described in Chapter 7, a porch roof also provides excellent design protection, allowing the use of fine, intricate woodwork beneath it with reduced need for maintenance. Wood exterior doors can benefit remarkably if they are placed within a protective alcove or entryway designed to reduce or eliminate their exposure to sun and rain. Awnings placed over windows and doors not only offer excellent protection against excessive radiant gain within the building, but they also lengthen the life of the exterior woodwork as well.

## Pitch for Runoff

Many exterior woodwork elements can benefit immensely from the inclusion of an angle or pitch to promote rapid runoff and removal of accumulating water. Both the superstructure beneath the woodwork and the woodwork itself can gain significant built-in protection against detrimental short- and long-term effects of water infiltration if the simple relationship between gravity and water flow is respected.

Typical examples of pitched wood members are illustrated in Figure 17-1, including windowsill and head trim, porch flooring, and various drip caps. It is also important to recognize the need to halt the tendency of water to flow by capillary action as well as in response to gravity; hence the common use of drip grooves on the underside of certain wood members. These small but effective milled details break the cohesive water film and cause water to drip clear rather than seep in.

## Water-Eliminating Joints

One of the most effective ways to ensure that woodwork and the structures beneath it are preserved is to design joints that, by their nature, exclude and eliminate water. Joints in exterior woodwork which allow water to seep in and accumulate can be the cause of severe damage. Most wood, if allowed to remain damp, readily becomes home to a host of destructive microorganisms and insects. And all wood, whether decay resistant or not, will suffer the ill effects of exaggerated wood movement if it is repeatedly wetted.

Some examples of water-eliminating joints are illustrated in Figure 17-2. The most common and obvious is the venerable overlapping joint, which works in a simple manner to exclude water with gravity. Most horizontal sidings utilize the overlapping joint, as do most types of flashing. In fact, nearly all woodwork joints that naturally eliminate water do so with a form of overlapping construction. Vertical siding, however, while it does overlap itself, uses vertical joints and will not necessarily exclude water. Certain wood joints cannot benefit from effective horizontal overlapping construction; inside and outside corners of wood siding, for example, often are assemblages of long vertical joints; these are best protected with a layer of flashing beneath, supplemented with applied protection (caulking) where the siding meets the corner.

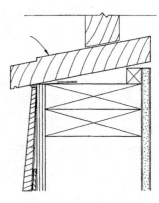
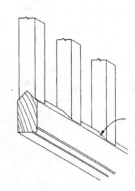

**Figure 17-1**
*Pitched or sloped elements.*

## Flashings

Often a joint between two wood members or between wood and another material (roofing or masonry, for instance) cannot be easily constructed to naturally exclude water, and must be sealed or protected by the addition of an impervious sheet or membrane known as flashing. Various materials serve as effective flashings, including sheet metal (steel, copper, zinc, aluminum), asphalt felts, and certain rubber or plastic compounds. A good flashing must be waterproof, and it must have the ability to be formed into complex shapes in order to fit within a joint unobtrusively.

CHAPTER 17

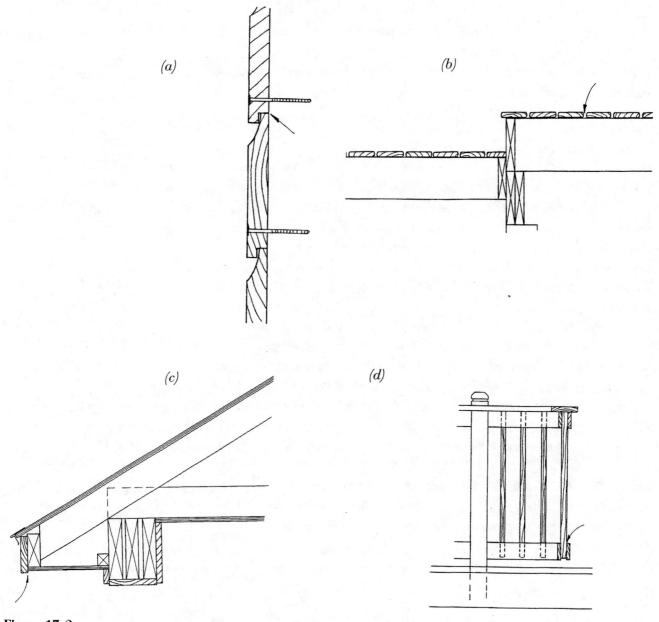

**Figure 17-2**
*Water-eliminating joints.*

Figure 17-3 illustrates the use of flashing in several exterior woodwork assemblies. When metal flashing is specified, fasteners (generally nails) should be of a like metal to prevent destructive galvanic action. If the nails will never be subjected to wetting, then the chance of rapid corrosion is unlikely, and metals may be mixed.

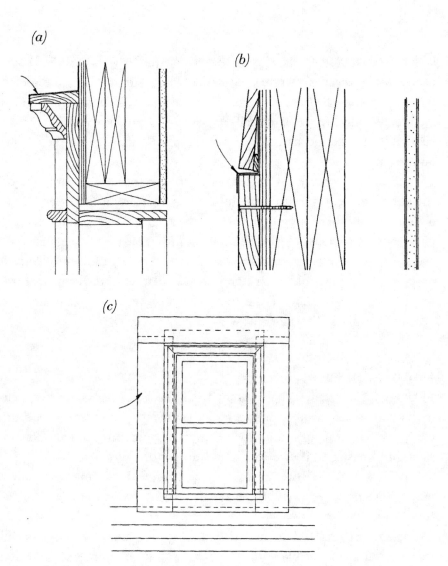

**Figure 17-3**
*Flashings.*

## Resistant Materials

If design parameters prevent the inclusion of certain active and passive protective elements, then the use of wood materials that protect themselves from within is a logical choice of action. Cedar, redwood, cypress, mahogany, and teak are all commonly utilized in exposed applications, because they contain natural extractives that make them resistant to decay and attack. Additionally, nonresistant woods can be treated with various preservatives to become resistant, even to the point of allowing direct burial in wet soil.

The obvious use for resistant wood is in totally exposed, unfinished applications like decks, fences, and docks. Its use is also quite prudent, however, in a variety of semiexposed situations. As siding, cedar and redwood will provide good service with no applied finish at all. And if a finish is used, it will further prolong the life of the installation. The material's resistant nature will prevent tricky areas—such as where siding meets the roof or receives splashback from rainfall, or permeable cut ends—from failure. Exterior woodwork that is initially expensive to produce and/or install is best built of resistant materials as a measure of insurance on the original investment.

Superior renovations or repairs to the exterior woodwork of older structures can be made with resistant wood, particularly if the superstructure design prevents the simple elimination of the source of rot. A common example is roofline trim on old wood buildings susceptible to ice damming at the eaves. Proper ventilation and insulation remedies are often prohibitively expensive in complex older structures; the installation of resistant wood will certainly prolong the life of the trim elements, even if they are subject to occasional water and ice stress.

## Landscaping

Although it's not really a construction detail, the use of certain landscape elements can, however, add or detract from the longevity of exterior wood. The effects of trees, plantings, and ground covers in close proximity to a building should be considered if exterior wood is to perform well.

Plants can cause damage to exterior woodwork by direct and indirect means. Directly, a tree or shrub can abrade wood or wood finishes when buffeted by winds, and it can provide a convenient pathway for destructive insects and vermin. Deciduous trees can fill gutters and other crevices with leaves, creating undesirable or damaging conditions. Indirectly, plants tend to concentrate moisture in their vicinity by shading and by transpiration. Woodwork that is closely surrounded by shrubs or trees will dry out slowly after a rain; it may remain damp for prolonged periods, increasing its susceptibility to invasion by fungi and insects. It is a good practice to place and prune plants far enough away from exterior woodwork that good air circulation and drying conditions exist.

Some landscape elements are favorable to the preservation of exterior woodwork. Trees, kept at a distance, can block harmful ultraviolet radiation, as well as slow the velocity of winds. Clean textural ground covers, either low plants or stone or wood mulch, can diffuse the energy of rainfall and runoff, reducing the amount of water splashed onto a building from the ground.

## Applied Protection

Good exterior woodwork design makes use of as many passive protective features as possible. Generally, though, for aesthetic reasons and to provide a more complete barrier, most exterior wood receives some form of active, or applied, protection in the form of paints, stains, sealers, and caulks. Coatings technology is diverse and extensive, and the painter's trade is similarly complex. This section discusses basic concerns and details only.

## Paint

The most common form of active wood protection is paint, a mixture of pigments and binders in a vehicle (the liquid solvent that makes application possible). Paint is applied as an opaque surface coating. On bare wood, a primer paint, with good bonding and recoating qualities, prepares the surface to receive additional coats of finish paint. Oil-based primers, which dry slowly and therefore penetrate deeper, are considered ideal. Finish paint excels in its weather resistance, and may contain mildewcides and ultraviolet-blocking agents. Its vehicle may be oil or water. Water-based acrylic-latex paints have excellent protective qualities, and their already broad use will only increase as stringent volatile organic compound (VOC) regulations become widespread.

Nearly all exterior wood of any substantial width should be painted on all sides prior to installation. This "back-priming" step will reduce the tendency of the wood to deform under uneven moisture stress, and it will eliminate the likelihood of any bare wood ever revealing itself as exterior components move in response to seasonal moisture variations. In addition, prepriming or prepainting is generally cost effective, since it can be performed efficiently at ground level or indoors without ladders, scaffolding, and concern for inclement weather. In most cases, though, a thorough job requires a fair bit of filling and caulking after wood installation, which should be followed by a top coat of paint applied in place.

## Stain

Exterior oil stain is a penetrant: its body and vehicle characteristics cause it to seep deeply into the pores and fibers of wood, carrying with it pigments for color and additives to prolong the wood's life. Many exterior stains are available in very thin, "semitransparent" formulations, which deposit enough pigment to color the wood, yet still allow some wood characteristics to show through. Heavier, "solid color" formulations mask all wood characteristics except texture. The difference between stains and paints is largely in their pigment content; too many coats of stain applied too frequently will cause enough build-up that it behaves like a surface coating or paint. Latex stains are similar to thin paints, and they dry too quickly to penetrate like their oil-based counterparts. They should be used primarily as a top coat over an oil base coat.

In order for stains to perform at their best, without the drawbacks of paints, they should be applied to rough textured wood that allows excellent penetration and deep coverage. When stains, with their light-bodied characteristics, are applied to smooth, new lumber, penetration is minimal. Semitransparent stains will weather rapidly, and solid color stains may sit on the surface like paint. It is important to determine finishing specifications prior to the installation of wood siding and trim, so that an ideal combination of wood texture and finish may be used.

## Sealers

Nonpigmented finishes, either coatings or penetrants, are termed sealers. They remain relatively clear, allowing the natural color of the wood to show. If they contain sufficient ultraviolet blocking agents, they will also preserve the wood's color, prolonging its new look. Since they contain no protective pigments, sealers typically have a shorter effective life span than paints; their recoating requirements are similar to semitransparent stains (every two to three years). Penetrating sealers should be applied heavily and often. Some surface sealers are somewhat unproven and should be used with caution.

## Caulks and Sealants

Caulks and sealants, resilient compounds that are applied in the form of a semiliquid and cure to an elastic state, have three basic uses for exterior protection: 1) as primary sealing materials used to exclude water; 2) as fillers, to neaten and enhance woodwork by concealing unsightly gaps; and 3) as air infiltration barriers, to reduce energy loss. Often a single bead of caulk, strategically placed, can perform all three functions.

# Exterior Protection

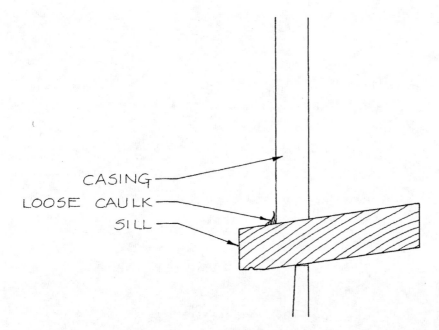

**Figure 17-4**
*Loosened caulking may trap water and render a sloped joint ineffective.*

The use of caulks as primary sealing materials should be avoided whenever a design or built-in solution is possible. Permanent protection as part of the woodwork construction, in the form of flashings and overlapping joints, is a preferred solution to the injection of a short-lived caulk, regardless of its guarantee. Some woodwork demands applied protection, though, and high-quality caulks, carefully applied, will effectively exclude water, and reduce the need for frequent maintenance.

As fillers, particularly prior to a thorough painting, paintable caulks and sealants are unsurpassed in their ability to close linear crevices and small construction gaps, and they can go a long way toward preventing air, water, and pests from finding their way in.

Lastly, caulks and sealants are often used as barriers to energy loss, reducing the amount of air movement through exterior surfaces. Overuse of caulks in the name of energy savings should be avoided in certain instances, however. Aside from the indoor air quality problems that may arise from oversealing, the application of caulk to ill-suited joints may do more harm than good. Inappropriately applied, caulking may actually *invite* water infiltration. An example is illustrated in Figure 17-4.

# GLOSSARY

**Active**  In a double or French door, the door panel configured for most frequent use.

**Adhesion**  The tendency of dissimilar materials to bond together.

**Adhesives**  Any of various bonding agents or glues that promote adhesion.

**Aesthetic**  Of or pertaining to the perception of art and beauty.

**Anodize**  To coat with a protective (often colored) oxide by electrolysis, with the base metal serving as the anode.

**Arc**  Any part of a curve, especially of a circle.

**Architectural woodwork**  Nonstructural, often decorative woodwork that is a permanent part of a building. Examples include siding, windows, and crown mouldings.

**Architrave**  In a classical order, the lowest part of the entablature; often a beamlike form, resting directly on a column capital.

**Astragal**  An often convex moulding used to cover a joint between adjacent windows or doors.

**Back priming**  The process of sealing the back, or hidden, face of wood siding or trim in order to provide maximum protection against moisture-related movement or decay.

**Balcony**  A projecting elevated platform, usually enclosed by a railing or low wall.

**Baluster**  A small post or spindle that supports a handrail, typically arranged in a repeating fashion.

**Balustrade**  A railing and its supporting balusters.

**Bank**  An arranged row or tier of objects. Examples may be windows or light fixtures.

**Base cap**  A small moulding that forms the uppermost member of a multiple-piece baseboard.

**Baseboard**  A trim piece, sometimes moulded, used to cover the joint between a wall and floor.

**Bead**  A moulded form, semicircular in cross-section view.

**Beam**  A horizontal structural member, usually supported at its ends by columns.

**Bevel**  To cut or form at an angle other than a right angle.

**Bifold**  A configuration of multiple doors, whereby one door folds against another in a pair.

**Blind nail**  A nail that is hidden by the milled configuration of the woodwork which it is securing. An example is seen in tongue-and-groove flooring.

**Blocking**  In wood framing, any auxiliary pieces of wood added to the rough framework to provide support for finishes or other work to be applied later. Typical applications are blocks to support closet rods and kitchen cabinets.

**Board**  The typical form of sawn wood, relatively long and flat.

**Buck**  A frame, usually supporting a swinging door, often of steel.

**Bypass**  A configuration of multiple doors, whereby one door slides past another.

**Cabinet**  An enclosed case or cupboard, typically fitted with shelves or drawers for storage.

**Casement**  A window that swings open on hinges along the side.

**Casework**  Cabinetry, or the woodwork of cabinets.

**Casing**  A trim piece, often moulded, used to close the joint between a door or window frame and the interior or exterior wall surface.

**Caulk**  A resilient sealing putty.

**Ceiling**  The topmost part or covering of a room or space.

**Chair rail**  A horizontal trim element, applied to a wall at the approximate height of the back of a chair, typically to protect the wall surface from wear.

**Circle**  A closed curved figure, bounded by a line that is a constant distance from a single point (the center). One of the four conic sections, a circle is formed when a cone is cut by a plane at a right angle to the cone's axis.

**Cladding**  A covering layer, often protective. An example is the thin layer of aluminum or vinyl on clad wood window frames.

**Clapboard**  A traditional term for beveled siding.

**Classical**  Pertaining to or relating to the principles of ancient Greek and Roman cultures.

**Clear**  In wood, free of knots and certain other defects. The term "clear" may be used in a general sense to describe a piece of wood, or as a definite category in a grading system that groups wood according to the presence of defined defects.

**Closet**  A storage room, often fitted with shelves or racks to hold clothing.

**Coffered**  Decorated with sunken or recessed panels; typically describes a ceiling or dome.

**Cohesion**  The tendency of similar materials to bond together.

**Colonnade**  A series of regularly spaced columns or posts, typically supporting a roof or beam.

**Column**  An upright, cylindrical structure; usually acting as a supporting element.

**Commercial**  Describes a construction or building used primarily for trade or commerce, such as an office or retail store (in contrast to a residential, industrial, or institutional site).

**Contemporary**  Pertaining to the modern or recent.

**Core**  The inner layer or layers of a multi-ply panel or door.

**Corner block**  A decorative block or piece, often with a lathe-turned face, used when casings meet at a corner around a window or door; Victorian-era.

**Cornice**  The uppermost part of a classical entablature, traditionally used to describe roofline overhang trim at the eaves.

**Court**  A typically uncovered outdoor area surrounded by buildings or walls.

**Crown moulding**  Any of a multitude of mouldings used to trim elevated architectural elements: ceiling edges, column capitals.

**Deck**  A wooden floor; often used to describe floor structures that are outdoors and uncovered.

**Deformed-shank**  A type of nail whose shank or shaft has been formed into a twisted, roughened, or ringed shape to increase withdrawal resistance.

**Design**  A contrived plan or layout; or the act of planning, usually skillful.

**Detail**  A small or particular part of a larger structure or assembly.

**Diffusion**  The scattering or intermingling of particles or molecules.

**Dimensional stability**  The tendency of an object to hold its dimensions (width, length) constant. Wood with a high level of dimensional stability is prized.

**Door**  A moveable panel or structure used to close an entrance to a building or a room.

**Double-hung**  A type of window with two sashes hung or mounted to slide up and down within a frame.

**Dowel**  A cylindrical rod. Wood dowels are often used to make or reinforce joints in wood millwork or furniture.

**Drip edge**  A metal flashing installed at the lowermost edge of a roof; it promotes proper elimination of rain.

**Drywall**  A surface finishing system that uses large paper-faced gypsum-core panels, with joints typically filled and leveled with a gypsum-based joint compound.

**Eaves**  The lowermost edge of a sloping roof, usually projecting from the building.

**Elemental**  Basic or primary; elemental building components are those parts that form the very essential structural form of a building.

**Ellipse**  A closed curved figure bounded by a line that is a constant total distance from two points called the foci. One of the four conic sections, an ellipse is formed when a cone is cut by a plane less steeply inclined than the side of the cone.

**Emissivity**  A description of the ability of a surface to radiate energy. Window glass with relatively low emmissivity (low-E) can offer improved energy performance as compared to clear glass.

**End grain**  The exposed face of wood cut across its axis of growth; usually more porous than other faces of wood.

**Engineered**  Used to describe wood building materials. Engineered wood products have been designed and tested to meet specific, consistent performance parameters. Examples are wood trusses, certain structural plywood, and laminated beams.

**Entablature**  The horizontal portions of a classical order, supported by columns, and composed of an architrave, frieze, and cornice.

**Ergonomic**  Of or relating to the study of the physical relationship of the human body to its environment, especially in work areas.

**European**  Describes a general class of cabinetry, originally popularized in Europe, typically exhibiting clean lines and simple carcass construction. The same as "frameless."

# Glossary

**Expansion (moisture)** The tendency of hygroscopic (water-attracting) materials to grow in overall dimension as a result of taking on water. Wood exhibits this tendency distinctly.

**Expansion (thermal)** The tendency of materials to grow in overall dimension in response to heat gain. Many common building materials exhibit this tendency to varying degrees. Examples are metals, plastics, and masonry.

**Expansion joint** A construction joint designed to account for the expansion of building materials.

**Exposure** The portions of exterior materials that are actually exposed to the weather; typically describes overlapping siding and roofing materials.

**Extractive** A substance that may be transferred or removed from a material, such as sap or resin from wood.

**Extruded** Formed or shaped by drawing forcibly through a die or hole.

**Face frame** A cabinetry component consisting of a flat framework of wood mounted to the forward-facing edges of a cabinet.

**Fascia** A component of a classical entablature; a horizontal band at the edge of a roof.

**Fenestration** The arrangement of windows in a building.

**Fiberglass plastic** A moulded plastic composed of resins reinforced with thin glass strands.

**Field/edge analysis** The division of a system or construction into two basic components: the field, or main essence, and the edge, where the field ends.

**Figure** Describes wood grain as it appears on the face of a board; the figure may be bland, beautiful, etc.

**Finger joint** A machined woodworking joint composed of interlocking glued fingers. The fingers provide added gluing surfaces to the joint. This technique is widely used to manufacture longer lengths of lumber from short, otherwise unusable scraps, and it is commonly used in hidden or painted work.

**Finish** The final presented surface of woodwork; the finish is a result of surface preparation (sanding, scraping, machining) and the application of surface protectors or enhancers (stains, paints).

**Fireplace** The area intended for open burning of wood, inside a building; composed of a firebox, chimney, and hearth, all constructed of noncombustible materials.

**Flashing**  Sheet materials used to weatherproof the joints where exterior materials change type or change direction. Traditionally metal, some flashings may be bituminous or plastic.

**Floor**  The bottom surface of a room.

**Frame**  A skeletal or supporting structure, often bordering the supported material, such as a door frame.

**Frameless**  A type of cabinetry construction that does not employ a face frame to finish or support the forward-facing edges of the cabinet's side, top, and bottom panels. The edges are instead typically trimmed with a thin band of veneer or plastic laminate.

**French door**  A double glass door, with each panel hinged to meet and swing open in the middle.

**Furring**  Nonstructural wood strips installed to build out a surface (wall, ceiling) and to support finish materials.

**Gable**  The triangular exterior wall formed beneath the slope of a roof.

**Galvanic reaction**  A destructive electrochemical reaction between two dissimilar metals, such as copper and zinc.

**Galvanization (electro-)**  To coat (steel, e.g.) with a thin layer of zinc by electroplating, to hinder oxidation.

**Galvanization (hot-dipped)**  To coat with zinc by simple immersion into molten zinc, to hinder oxidation.

**Glass**  A brittle colloid of silica, usually transparent or translucent, widely and almost exclusively used for window panes.

**Glazing**  Window glass that has been or will be set into a frame or sash.

**Grade/grading**  To organize, by a defined set of established criteria, an inherently variable material or product. Wood is generally graded for structural applications (framing lumber), for decorative applications (siding, cabinet woods), for manufacturing (shop lumber), and so on. Typical criteria used include species and the presence of defects (knots, short grain, wane).

**Grain**  The variation of wood fibers caused by seasonal growth. Wood formed under rapid growing conditions (spring wood) will vary in color and density from wood grown more slowly (summer wood). These variations are visually apparent as lines or stripes in sawn wood, and their form will change with a change in cut angle.

**Gutter**  A trough to catch and direct rainwater, placed at the edge of a roof.

# Glossary

**Hardboard**  A manufactured board or panel formed by subjecting wood fibers to heat and pressure.

**Hardware**  Fittings, such as handles, screws, knobs, locks; typically of metal.

**Hardwood**  A broad class of wood from any tree classified as an angiosperm and possessing true vessels. Hardwoods are typically dense and heavy, but not necessarily so. Examples are oak, maple, and basswood.

**Head**  Pertaining to the top or upper portion. *Head* casing trims the *top* of a window or door.

**Header**  A structural framing element that spans a wall opening or supports the ends of joists, and so on.

**Hearth**  The floor of a fireplace, composed of an inner and outer portion.

**Hinge**  A hardware device on which a door, gate, and so on swings.

**Housing**  A wood joint where one piece is inserted into another; typical in stair construction.

**Industrial**  Describes a construction or building used primarily for manufacturing or similar activities, such as a factory or workshop; in contrast to a residential, commercial, or institutional site.

**Infiltration**  The seepage or invasion of air or water into a space through cracks or loose joints.

**Institutional**  Describes a construction or building used primarily for public activities, such as a school or a hospital; in contrast to a residential, commercial, or industrial site.

**Jamb**  The frame piece, typically the upright side member, to which a window or door is hung or mounted.

**Joist**  The horizontal parallel framing member that supports a floor or ceiling.

**Kerf**  The slot formed when a board is sawn.

**Lamina**  An individual layer in a multilayer construction, such as a ply of plywood or a layered board in a laminated beam.

**Laminar**  Having a layered or laminated character.

**Landing**  A small platform or floor at the end of a flight of stairs.

**Level**  Perfectly horizontal; perpendicular to the pull of gravity.

**Life-cycle cost**  The extended cost of a fixture or construction calculated over its useful life span.

**Low-E** Low emissivity. A term commonly used to describe the ability of window glass to pass radiation in a manner conducive to good energy performance and minimal textile fading.

**Mantel** A horizontal, decorative construction typically installed over a fireplace and to a lesser extent over door and window openings. Mantels are often duplications of classical entablatures.

**Masonry** The materials or trade associated with installing brick, stone, and concrete.

**MDF** Medium-density fiberboard. A wood fiber panel with a dense, smooth character; it is versatile and is commonly and used for painted interior constructions: panels, trim, and so on.

**Metal** Any of the elements (and their alloys) characterized by malleability, ductility, luster, and conductivity; examples are copper, steel, and aluminum.

**Modular** Composed of standardized modules or components designed to come together as a whole.

**Moisture content** The content of water in wood, typically expressed as a percentage of oven-dried weight.

**Mortise** A slotlike recess in wood, into which slides a mating tenon to form a mortise-and-tenon joint. The joint may be through or blind, and it may be strengthened with wedges or pins.

**Motif** A main element, idea, or theme.

**Moulding** Wood or other material that has been formed into a decorative shape. Examples are crown moulding and fluted casing.

**Mullion** A dividing bar or trim between windows or panels.

**Muntin** The thin strip of wood or metal that divides and supports glass panes in a multi-pane window or glass door. False muntins, in the form of applied grilles or bars over a solid large glass pane, are widely used for decorative purposes.

**Nailbase** A substrate that accepts and holds nails well, such as plywood or solid wood.

**Newel** The post at the top and/or bottom of a flight of stairs which supports the handrail.

**Oil-based** Paint or stain with an oil vehicle.

**Order (classical)** Any of the classical (ancient Greek and Roman) structural styles, defined principally by the moulded forms of the supporting columns and the supported entablature. Examples are the Doric, Ionic, and Corinthian orders.

**OSB**  Oriented strand board. A sheathing panel formed of wood wafers glued in roughly opposing layers of prevailing grain direction.

**Oxidation**  A chemical process resulting in the removal of electrons (negative charge); commonly seen in metals as rust or corrosion from exposure to the atmosphere.

**Paint**  An opaque protective and decorative surface coating, consisting of a solid pigment suspended in a fluid base or vehicle.

**Panel**  A flat, rectangular piece forming a wall or door surface, typically mounted in a frame; it may be flush, raised, or recessed.

**Parquetry**  A geometric assemblage of small pieces of wood to form a floor.

**Particleboard**  A dense manufactured panel consisting of small wood chips and bonded with adhesive under pressure.

**Patio**  A courtyard or paved area adjacent to a house.

**Pedestal**  The foot or bottom support of a column, upon which the column base rests.

**Piazza**  A large, usually covered, courtyard or porch.

**Pier**  A column, typically supporting an arch or beam.

**Pilaster**  A structural or decorative support or pier attached to and projecting from the surface of a wall. A pilaster may be treated architecturally as a column with a base, shaft, and capital.

**Pitch**  Another term for slope or inclination; often used to describe roofs.

**Plain-sawn**  A common method of sawing a log into boards, where the plane of the boards is roughly perpendicular to the log's diameter.

**Plank**  A long, broad, thick board.

**Plank flooring**  Wooden floorboards that are wider than strip flooring (three inches).

**Plaster**  A hard wall and ceiling finish made of a mixture of lime or gypsum, sand, and water; applied as a paste to a suitable substrate with a trowel, in single or multiple layers.

**Plastic**  Any of a large variety of moulded or mouldable organic polymers or compounds. Examples are acetal, acrylic, and polycarbonate plastics.

**Plate (biscuit)**  A compressed wood joining plate, oblong in shape, which is inserted with glue into a matching slot to make a reinforced wood joint.

**Plating** A coating or layering of a decorative or protective metal film onto the surface of another metal. Plating of brass, zinc, or chrome onto steel is common in building hardware.

**Plinth** A base or low block of trim on a column, wall, or door casing.

**Plinth block** A shaped block placed where a door casing meets the floor, to represent a true plinth, except in low relief.

**Plumb** Perfectly vertical; parallel to the pull of gravity.

**Plywood** A manufactured panel or sheet good made of thin veneers glued to a core. The core may be of additional veneers or may be a thick layer of a substrate such as particleboard or MDF.

**Pocket** A door style where the door panel slides into a recess within the wall.

**Porch** An attached, covered outdoor structure built at or around an entrance.

**Portico** Similar to a porch, but usually with a colonnade.

**Post** A long cylindrical or square upright support, typically of wood or metal.

**Primer** A paint or sealer intended as a preparatory or base layer of a multicoat finish.

**Projected** A window style where the sash, when opened, projects from the wall plane.

**Quartersawn** A less common method of sawing a log into boards, where the plane of the board is parallel to the log's radius.

**Rabbet** A recess or groove milled into the edge of a board to form a variety of joints.

**Rafter** The sloping parallel frame members that support a roof.

**Rail** A horizontal frame member, as of a door frame.

**Rake** Slope or inclination.

**Relief** A cut or channel milled onto the back of a board or moulding to relieve internal stresses and/or to allow the piece to be installed flat with good edge contact.

**Residential** Describes a construction or building used primarily as a place for living, such as a house or apartment; in contrast to a commercial, industrial, or institutional site.

**Resin** A viscous organic solid or semisolid; extracted from certain plants or trees, or synthesized. Used in varnish, lacquer, and so on.

**Rift-sawn** A method of sawing a log into boards, where the plane of the boards is at a sloping angle to the radius of the log.

**Rise** The amount of vertical gain; describes stair and roof dimensions.

**Riser** The upright piece of a staircase, located beneath the front edge of the tread.

**Roof** The assembly of components constructed to cover and protect the top of a building.

**Rosette** A carved or moulded ornament with the general shape of a flower's petals radiating from the center.

**Row** A horizontal arrangement of objects.

**Run** The amount of horizontal distance; describes stair and roof dimensions.

**Sash** The moveable portion of a window, consisting of glazing within an operable frame.

**Sealant** A resilient crevice or gap filler.

**Sealer** A coating or penetrant used to close the surface of a porous material.

**Semitransparent** Describes exterior stains with a pigment content somewhat less than that of paint; they do not fully mask the color of the wood.

**Shadow line** The dark linear shadow formed from overhanging or overlapping materials such as clapboard siding.

**Shingle** A thin piece of wood, slate, or composition material laid in an overlapping fashion with other pieces to form a protective covering for roofs and exterior walls.

**Shutter** A hinged door, typically mounted in pairs at each side of a window. Originally used to seal windows during severe weather, now used largely for decoration in a perpetually opened position.

**Siding** Any of various milled or formed wood, metal, or other materials designed and installed to cover and protect exterior surfaces.

**Sill** The lower member of a window or exterior door frame; typically slopes down toward the outside to expel water.

**Single-hung** A two-sash window with one sash hung or mounted to slide up and down within a frame, and the other sash fixed.

**Skin (door)** A thin plywood, mounted to the core of a flush-type door.

**Skirt** The closure beneath a porch floor.

**Soffit** The horizontal underside of an overhanging structure, such as a cornice.

**Softwood**  A broad class of wood from any tree classified as a gymnosperm and not possessing true vessels. Softwoods are typically light and yielding, but not necessarily so. Examples are pine, fir, and cedar.

**Specification**  A written or otherwise communicated detailed description of a material, assembly, or procedure. Together with plans, specifications define and describe a construction for the appropriate parties.

**Spindle**  A turned baluster, typically used to support a handrail.

**Spline**  A flat key or strip that fits within mating grooves of pieces to be joined.

**Stack**  A vertical arrangement of objects.

**Stain**  A thin, penetrating, pigmented dye solution, used to color or darken wood.

**Stave**  A shaped wood strip used to form the side of a barrel, bucket, or cylindrical column.

**Stile**  A vertical frame member.

**Stock (manufacturing)**  Typical, run-of-the-mill, readily available.

**Stool**  The lower trim piece of an interior window frame; usually a projection of the sill.

**Stoop**  A small platform or porch with steps at the entrance to a house.

**Stringer**  The sloping piece of a staircase; may be rough (hidden) or finished (exposed). Exposed stringers may be open or closed.

**Strip flooring**  Wood flooring, usually tongue-and-groove, of approximately three inches or less in width.

**Stud**  An upright frame member, arranged in rows to form walls.

**Tempered**  Treated with heat to attain desirable levels of hardness and strength; typically describes glass and metal.

**Tenon**  A cut protrusion that slides into a mating hole called a mortise to form a mortise-and-tenon joint. The joint may be through or blind, and it may be strengthened with wedges or pins.

**Terrace**  A flat, protruding balcony or portico, usually covered.

**Thermal break**  A material that stems the flow of heat; metal windows and doors are typically broken with a strip of plastic or wood for energy efficiency and control of condensation.

**Toe kick**  The board or trim beneath a cabinet which forms the back of the toe space.

# Glossary

**Toe space**  The recess beneath a cabinet which provides clearance for the foot.

**Tongue-and-groove**  An interlocking joint made up of a tongue which is inserted into a mating groove.

**Traditional**  Pertaining to the styles or methods handed down from the past.

**Tread**  The horizontal part of a stair, which is stepped on.

**Trammel**  An instrument for striking curves.

**Trim**  Any of a large variety of decorative and sometimes protective strips of board or moulding used to close joints, especially where materials change type or direction; such as the joint between a window frame and a wall.

**Ultraviolet radiation**  That portion of the electromagnetic spectrum just beyond violet light (5 to 400 nanometers); it can contribute to material degradation and fading.

**Vault**  An arched or ascending ceiling or roof.

**Veneer**  A thin slice of wood, typically of a costly species or a rare grain pattern, which is mounted to a less expensive base or core.

**Ventilate**  To provide with an exchange of air.

**Veranda**  An open, covered porch or portico along the side of a building.

**Vinyl**  A plastic formed by the polymerization of vinyl resins; used widely in many formulations.

**Wainscoting**  Wood paneling or covering on interior walls, especially at the lower part.

**Wall**  The sides of a room or building.

**Weather**  To degrade or wear as a result of exposure to the elements.

**Weatherstrip**  A thin linear material installed at the perimeters of openings (doors, windows) as a seal against air and water intrusion.

**Wedge**  A tapered piece.

**Window**  A glazed opening in a wall.

**Wood**  The solid cellulose product of trees.

**Z-flashing**  A formed metal flashing in the shape of a squared "z"; used to divert water out from above a head casing or other square-edged piece.

# APPENDIX 1 | Wood Mouldings

Wood mouldings are specified by referring to their pattern, which may be a numerical standard such as WM74 or a descriptive name such as 3¼ inch Colonial Casing. The former is an effective practice when industry standard profiles are desired; the latter is open to broader interpretation and should be supplemented with a drawing of the intended pattern.

Custom milling can be a sound choice for distinctive work or when a nonstandard wood species is required. In some areas, for example, woodwork to be painted may be custom-run in poplar at less cost than stock clear pine mouldings. Run-to-order patterns are specified by referring to a pattern selection at the millworks, or to an original design. In many cases a knife is ground to the correct profile for use in a moulding machine. Care must be taken when designing custom moulding profiles; a good understanding of the milling process and its limitations is necessary. Often, available standard mouldings, used in creative combinations or applications, will produce desirable results at lower cost.

## Selected Patterns

The pattern profiles illustrated here are a selected sampling of the literally hundreds of possible styles and sizes available as stock or custom mouldings. They show typical shapes and sizes and are presented as an overview of wood moulding profiles and their uses.

## APPENDIX 1

**Figure A1-1a**

*Casings. (a) ranch or contemporary, (b) standard colonial, (c) wide colonial, (d) flat, eased edge, (e) custom, fluted, (f) custom, raised center, (g) brickmould.*

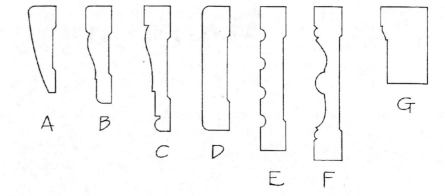

## Casings

Primary use: As edge trim around windows and doors, interior and exterior. Exterior patterns are typically heavier and simpler than interior—often just square-edged boards. Most of the patterns in Figure A1-1a are intended primarily for interior use. Other uses: Baseboards and low-relief wall frames.

## Baseboards

Primary use: As edge trim where walls and floors meet.

## Crowns, Beds, and Coves

Primary use: As edge trim where wall and ceiling or soffit meet; part(s) of cornice. Standard crown mouldings are typically "sprung"; they are milled so

**Figure A1-1b**

*Baseboards. (a) ranch or contemporary, (b) quirk beaded, (c) traditional, (d) colonial, (e) two-piece with base cap, (f) three-piece with base cap and base shoe.*

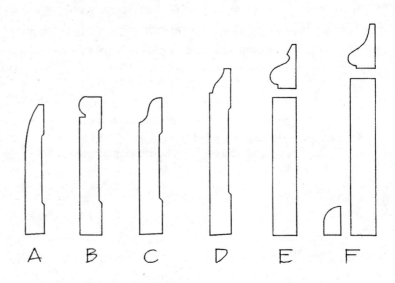

# Wood Mouldings

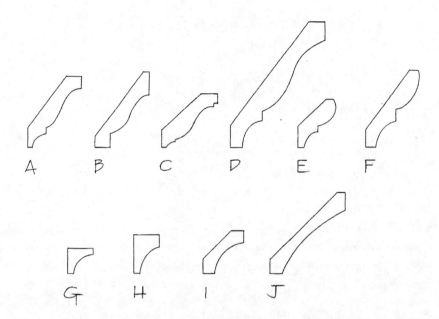

**Figure A1-1c**
*Crown, beds, and coves. (a–d) crown mouldings, (e, f) bed mouldings, (g–j) cove mouldings.*

their height is greater than their projection. Bed mouldings may be sprung or plain. Coves are always plain; they lay in a 90° corner symmetrically at 45°. Other uses: Casework and furniture ornamentation.

## Drip Caps

Primary use: Over exterior door and window frames as a water diverter/flashing; the milled drip groove breaks the water film, causing the water to drip clear of the construction below. Other uses: As a "water table" at or near the base of an exterior wall, to divert water away from the foundation.

**Figure A1-1d**
*Drip caps.*

## Astragals

Primary use: As a center stop attached to the passive door of an active/passive pair. Other uses: The flat type can serve as mullion trim or as panel moulding.

**Figure A1-1e**
*Astragals. (a, b) T-astragals, (c) flat astragal.*

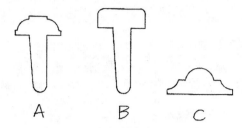

## Chair Rails
Primary use: As a horizontal wall decoration and/or protection, particularly from chair damage. Other uses: As rail-height stairwell wall protection.

## Stops
Primary use: In door construction, the stop keeps the door from swinging through the frame. Certain windows use stop moulding to guide the sash in the frame. Other uses: As shelf edging and multipurpose trim.

**Figure A1-1f**
*Chair rails.*

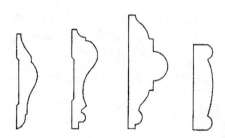

**Figure A1-1g**
*Stops. (a) ranch or contemporary, (b) square or plain, (c) colonial, (d) beaded.*

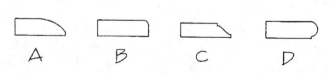

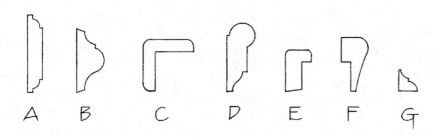

**Figure A-1h**
*(a) mullion trim; applied between adjoining windows, (b) panel moulding; applied to walls to frame false panels, (c) corner guard, (d) picture moulding; applied high on a wall to support suspended artwork, pictures, etc., (e) backband; applied around a casing to thicken its edge, (f) ply cap; applied over the top edge of thin plywood wainscoting, (g) glass bead or stop; used to secure glass in an opening.*

## Miscellaneous

For more information, contact:

Wood Moulding & Millwork Producers Association
P. O. Box 25278
Portland, OR 97225

# APPENDIX 2 | Wood Siding

Like mouldings, wood sidings are specified by referring to their pattern, which may be a numerical standard such as 354 or a descriptive name such as Rabbeted Bevel. The former is an effective practice when industry standard profiles are desired; the latter is open to broader interpretation and should be verified by a diagram or sample. In addition, size (width, thickness), species (redwood, pine, etc.), grade (clear, #1 common, etc.), and surfacing (surfaced, saw-textured, etc.) must be specified according to application and budget.

Wood sidings are produced according to standards set by several trade associations, depending on region and species. It is advisable to obtain and review trade association literature (see sources below) prior to the specification of any unusual wood sidings. An understanding of species, grades within species, regularly run patterns, and local preferences is useful, as each of these factors can affect material cost. Most suppliers stock only popular siding types and species; other types must be obtained by special arrangement.

*Wood & Wood Siding Specification Checklist* (see Appendix 3):

- use
- species
- grade and rules agency
- grain, where appropriate
- seasoning or moisture content
- pattern
- size
- texture of face

# APPENDIX 2

**Figure A2-1a**
*Bevel sidings horizontal application. (a, b) plain bevel, (c, d) rabbeted bevel, (e, f) rabbeted drop.*

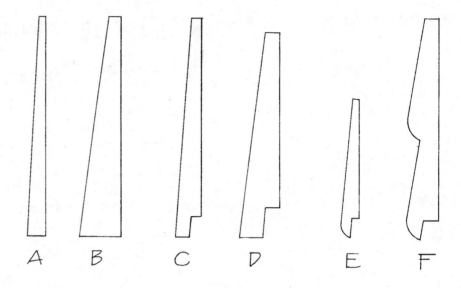

## Selected Patterns

The pattern profiles illustrated are a selection of the basic styles of milled wood sidings available.

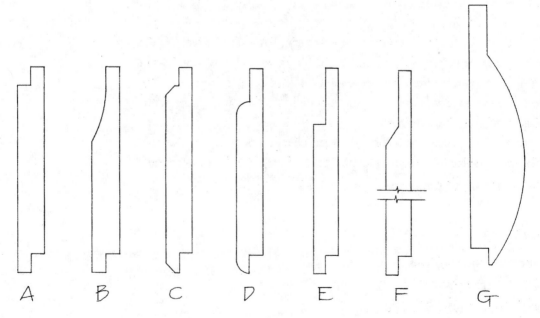

**Figure A2-1b**
*Shiplap sidings: horizontal (preferred) or vertical application. (a) shiplap, (b) cove shiplap, (c) V shiplap, (d) Boston shiplap, (e) channel shiplap, (f) channel shiplap, beveled channel, (g) log cabin.*

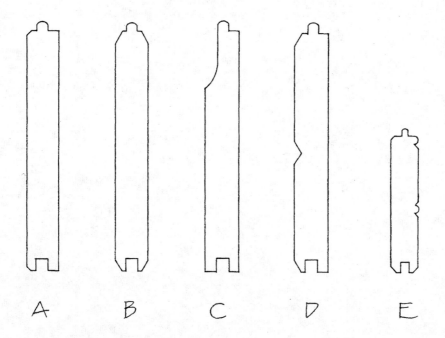

**Figure A2-1c**
*Tongue-and-Groove Sidings (T&G): horizontal (preferred) or vertical application. (a) T&G V, (b) T&G V (two-sided), (c) T&G drop, (d) T&G V & center V, (e) beaded (ceiling & wainscot).*


For more information, contact:

    Western Wood Products Association
    Yeon Building
    522 SW Fifth Ave.
    Portland, OR 97204-2212

    California Redwood Association
    405 Enfrente Drive, Suite 200
    Novato, CA 94949

# APPENDIX 3 | Wood Grades

Wood may be graded, or sorted into categories, according to a variety of criteria. The basic criteria for grading are based on the destined end use of the wood. The specific criteria within a particular grading system are based on wood properties relevant to that end use.

For example, wood in dimensional lumber form destined for structural use, like joists, studs, and rafters, can be graded according to the presence of defects that might affect its use as a structural material. Such defects are knots, splits, wane, or short grain. Boards are visually inspected and sorted into defined categories based on defects. A high grade, such as a #1, will contain a minimum of defects and will be correspondingly more valuable than a #3 grade board. Some structural lumber is even stress-graded; each individual piece is subjected to a standard measurable test by a machine, which then grades the lumber for its actual performance under bending conditions.

Most of the solid wood used for architectural woodwork is graded instead for its appearance. Appearance defects might be knots, splits, wane, and short grain, just like structural lumber. Yet, in addition, other appearance defects are considered that affect the look and often the function of the wood. Additional grading criteria might be the proportion of sapwood, the orientation of the grain in section (quartersawn, for example), and the amount of contrasting color variation. Wood veneers and plywoods are graded similarly as solid lumber, with additional criteria to consider, such as the type of veneer cut and the type of grain matching. Plywood and veneer grades are outlined in Appendix 4.

Wood grades are established by a number of rules writing agencies such as the California Redwood Association and the Western Wood Products Association. Member mills adhere to a set of established grading rules, and specifiers and end users in distant locales should be able to select wood within a grade with confidence that it will meet defined criteria. Wood grade has a direct and often dramatic impact on cost, and one must often find a difficult balance between cost and quality in order to build well and within budget.

# APPENDIX 3

Some widely used grading systems are outlined below. For the specific criteria used to define a grade, contact the agencies listed.

## Western Woods

Includes these groupings of softwood species:
- Douglas Fir–Larch
- Douglas Fir–South
- Hem–Fir
- Spruce–Pine–Fir
- Western Woods
- Western Cedars

Table A3-1 outlines grades for both boards and run-to-pattern products.

### Table A3-1 Appearance Lumber Grades

| Product | Grades |
|---|---|
| Selects (all species) | B & Btr Select<br>C Select<br>D select |
| Finish (usu. Douglas Fir & Hem-Fir only) | Superior<br>Prime<br>E |
| Special Western Red Cedar Pattern Grades | Clear Heart<br>A Grade<br>B Grade |
| Common Boards (primarily in pines, spruces, and cedars) | 1 Common<br>2 Common<br>3 Common<br>4 Common<br>5 Common |
| Alternate Boards (WCLIB Rules; primarily in Doug. Fir and Hem-Fir | Select Merchantable<br>Construction<br>Standard<br>Utility<br>Economy |
| Special Western Red Cedar Pattern Grades | Select Knotty<br>Quality Knotty |

## Redwood

Table A3-2 outlines redwood lumber grades.

### Table A3-2 Redwood Lumber Grades

| Class | Grades |
| --- | --- |
| Architectural | Clear All Heart<br>Clear<br>B Grade |
| Garden | Construction Heart<br>Construction Common<br>Merchantable Heart<br>Merchantable |

## Hardwood

Hardwood lumber is graded according to the proportion of clear or sound material that may be cut from a particular board. Each grade is limited by a minimum acceptable board size, and all grades are determined from the poor side of the piece, except when otherwise specified. Table A3-3 outlines hardwood lumber grades.

### Table A3-3 Hardwood Lumber Grades

| Grade | Portion<br>Clear or Sound | Minimum Board Size<br>Width (in.) × Length (ft.) |
| --- | --- | --- |
| FAS (formerly firsts & seconds) | 10/12 clear | 6 × 8 |
| FIF (FAS one face) | 10/12 clear (good side)<br>8/12 (poor side) | 6 × 8 |
| Select | 10/12 clear (good side)<br>8/12 clear (poor side) | 4 × 6 |
| 1 Common | 8/12 clear | 3 × 4 |
| 2A Common | 6/12 clear | 3 × 4 |
| 3A Common | 4/12 clear | 3 × 4 |
| 2B Common | 6/12 sound | 3 × 4 |
| 3B Common | 3/12 sound | 3 × 4 |

For more information, contact:

Western Wood Products Association
Yeon Building
522 SW Fifth Ave.
Portland, OR 97204-2122

California Redwood Association
405 Enfrente Drive, Suite 200
Novato, CA 94949

National Hardwood Lumber Association
P.O. Box 34518
Memphis, TN 38184-0518

# APPENDIX 4 | Plywood

As with lumber, plywood manufacture is governed by trade associations that establish standards and grading practices that are followed by member mills and shops. In the U.S., the primary softwood plywood and structural panel authority is the American Plywood Association (APA). The Hardwood Plywood & Veneer Association (HPVA) sets standards for decorative hardwood-faced sheet goods.

## Softwood Plywood

The softwood plywood industry is immense, supplying a large variety of panel products, including sheathing for floors, walls, and roofs; linings for concrete forms; underlayments for flooring; many specialty products; and siding.

The last category—siding—is of primary interest. Plywood sidings may be manufactured as conventional veneered plywood, as a composite with a nonveneer core, or as a nonveneer panel.

The APA rates certain of its products specifically as sidings. They are specified by grade and surface pattern. Surface patterns consist of various texture treatments, overlays, and grooving. The grades are based on the number and type of patches on the face side. By using patches, the face of an APA-rated siding does not have natural defects as other plywood panels do, such as knots, splits, etc. They are designed to take a stain or paint finish well and to present a uniform appearance. Fiber-resin overlaid panels (medium density overlay or MDO) have a smooth synthetic face that masks panel imperfections, natural or otherwise. Overlaid panels are designed to take a paint finish well, hence they are a panel of choice for signmakers. Grooves of various dimensions milled into the face of plywood siding panels can add visual interest.

## Siding Surface Patterns

### Texture 1–11
Rough-sawn, brushed, overlaid, or other texture; shiplapped edges; grooves $1/4$ in. deep by $3/8$ in. wide; 4 or 8 inches on-center (o.c.); other spacings to order. Douglas fir, redwood, southern pine, cedar, and other species.

### Rough-Sawn
Slight rough-sawn texture running across panel; with or without grooves; available in panel form, or as a lap siding. Douglas fir, redwood, cedar, southern pine, and other species.

### Com-Ply
Rough-sawn face; grooved; reconstituted wood fiber core, shiplapped. Douglas fir or cedar face.

### Medium Density Overlay
Overlaid face; with or without grooves; smooth or texture-embossed.

### Brushed
Brushed or relief-grain face for striking surface. Douglas fir, cedar, or other species.

### Channel Groove
A thin panel with shallow grooves; available in types similar to T1–11.

### Reverse Board-and-Batten
Textured surface with deep wide grooves 8 or 12 inches on-center. Redwood, Douglas fir, cedar, southern pine, and other species.

## Table A4-1 Rated Siding 303 Face Grades

| Class | Grade | Wood Patches | Synthetic Patches |
|---|---|---|---|
| Special Series 303 | 303-OC (clear) | Not permitted | Not permitted |
| | 303-OL (overlaid) | N/A for overlays | N/A for overlays |
| | 303-NR (natural rustic) | Not permitted | Not permitted |
| | 303-SR (synthetic rustic) | Not permitted | Permitted as natural defect shape only |
| 303-6 | 303-6-W | Limit 6 | Not permitted |
| | 303-6-S | Not permitted | Limit 6 |
| | 303-6-S/W | 6 either type | 6 either type |
| 303-18 | 303-18-W | Limit 18 | Not permitted |
| | 303-18-S | Not permitted | Limit 18 |
| | 303-18-S/W | 18 either type | 18 either type |
| 303-30 | 303-30-W | Limit 30 | Not permitted |
| | 303-30-S | Not permitted | Limit 30 |
| | 303-30-S/W | 30 either type | 30 either type |

## Grades

Plywood "grades" may refer to the *panel* grade or the *veneer* grade. Panel grades either describe the face and back veneers (A–C, B–C) or the panel's performance classification (Rated Sheathing, 303 Siding). Veneer grades define veneer appearance in terms of natural unrepaired characteristics and the allowable number and size of repairs made during manufacture. (See Table A4–1.)

## Veneer Grades

*A*  Smooth, paintable. Not more than eighteen neatly made repairs—boat, sled, or router type, parallel to the grain—permitted. May be used for natural finish in less demanding applications. Synthetic repairs permitted.

*B*  Solid surface. Shims, circular repair plugs and tight knots to 1 inch across the grain permitted. Some minor splits permitted. Synthetic repairs permitted.

*C*  C-Plugged. Improved C veneer with splits limited to $1/8$ inch width and knot holes and borer holes limited to $1/4 \times 1/2$ inch. Admits some broken grain. Synthetic repairs permitted.

C. Tight knots to 1½ inch. Knotholes to 1 inch across grain and some to 1½ inch if total width of knots and knotholes is within specified limits. Synthetic or wood repairs. Discoloration and sanding defects that do not impair strength permitted. Limited splits allowed. Stitching permitted.

**D** Knots and knotholes to 2½ inch across width of grain and ½ inch larger within specified limits. Limited splits allowed. Stitching permitted. Limited to Interior, Exposure 1, and Exposure 2 panels.

## Exposure Durability

APA plywood is rated according to its ability to perform in exposed situations. All siding grades are rated Exterior.

### *Exterior*

Exterior panels have a fully waterproof bond and are designed for applications subject to permanent exposure to the weather or to moisture.

### *Exposure 1*

Exposure 1 panels have a fully waterproof bond and are designed for applications where long construction delays may be expected prior to providing protection, or where high moisture conditions may be encountered in service. Exposure 1 panels are made with the same adhesives used in Exterior panels. However, because other compositional factors may affect bond performance, only Exterior panels should be used for permanent exposure to the weather. The trade term "CDX," used to denote the all-veneer APA Rated Sheathing, is an Exposure 1, not Exterior, panel.

### *Exposure 2*

Exposure 2 panels (which are identified as interior type with intermediate glue) are intended for protected construction applications where only moderate delays in providing protection from moisture may be expected.

### *Interior*

Interior panels are manufactured with interior glue and are intended for interior applications only.

## Hardwood Plywood

Hardwood plywood encompasses a large range of panel products manufactured primarily for appearance (nonstructural) use. Uses include cabinetry, furniture, sporting equipment, and musical instruments. The face veneers are usually of hardwood species, although high-quality decorative plywood with a pine face is still governed by the same standards as hardwood veneered panels. The core material may be of hardwood or softwood veneers, particleboard, medium-density fiberboard, hardboard, or other special cores. Unlike softwood plywood, which always uses rotary-sliced veneers, hardwood plywood is manufactured with its face veneers sliced and matched in a number of ways for a variety of grain and figure effects.

## Grades

Hardwood plywood is graded primarily on the visual quality of its face veneers. Panel grades consist of a letter and a number that describe a standard range of characteristics for the face and back veneers, respectively.

### Face Grades

*AA* The best-quality face grade for high-end use such as architectural paneling, doors and cabinets, case goods, and quality furniture.

*A* For use where AA is not required but excellent appearance is very important, as in cabinets and furniture.

*B* For use where natural characteristics and appearance of the species are desirable.

*C, D, and E* These grades provide sound surfaces, but they allow unlimited color variation; grades C, D, and E, respectively, allow repairs in increasing size ranges. Applications: where their surface will be hidden or a more natural appearance is desired.

### *Back Grades*

*1, 2, 3, and 4* Requirements of grade 1 are most restrictive, with grades 2, 3, and 4 being progressively less restrictive. Grades 1 and 2 provide sound surfaces with all openings in the veneer repaired except for vertical worm holes not larger than $1/16$ inch. Grades 3 and 4 permit some open defects; however, grade 3 can be obtained with repaired splits, joints, bark pockets, laps, and knotholes to achieve a sound surface if specified by the buyer. Grade 4 permits knotholes up to 4 inches in diameter and open splits and joints limited by width and length.

### *Veneer Inner-Ply Grades*

*J, K, L, and M* Grade J is the most restrictive, allowing minimal size openings. Grades K, L, and M are progressively less restrictive. The least restrictive grade (M) is usually reserved for plies not adjacent to faces; it allows round and similar shaped openings not to exceed $2\frac{1}{2}$ inches and elongated openings up to 1 inch to be visible on the edges or ends of panels.

## Veneer cutting

Grain pattern is affected by the way in which a veneer is cut from a log. The most common methods are rotary cutting and plain slicing. Rotary-cut veneers, made by turning the log against the knife like a lathe, produces wide sheets with wild, somewhat unnatural grain patterns. Plain-sliced veneers are sliced from the log in the same orientation that plain-sawn boards are ripped; these veneers have the natural grain appearance of true boards. Other less common methods are quarter-slicing (similar to quartersawing), rift-cutting (between plain and quartered, used in oak to minimize ray fleck), and half-round slicing.

## Veneer matching

Except for rotary cut veneers, all hardwood plywood veneers must be assembled into joined sheets of sufficient width to cover the entire panel face. This necessity opens up decorative opportunities—the art of veneer matching can produce a variety of effects, depending on how the veneer sheets are arranged.

**Figure A4-1**
*Veneer matching techiniques*

ROTARY

BOOK

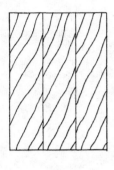
SLIP

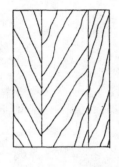
PLEASING

RANDOM

## Whole Piece Rotary Cut
Not a matching technique see Figure A4-1. One single piece of veneer is used with continuous grain characteristics running across the sheet.

## Book Matching
Alternating pieces of veneer from a flitch (a stack of veneer sheets in the order they came off the log) are turned over, so that adjacent pieces are "opened," like two pages in a book. Adjoining pieces may finish unevenly, since the slicing process leaves one side of a leaf with a loose face and one with a tight face.

## Slip Matching
Adjoining pieces are slipped out in sequence, with all the same-face sides being exposed.

## *Pleasing Match*
Veneers are matched by color similarity, not necessarily grain characteristics.

## *Random (or Mismatch)*
Random selection of leaves from one or more flitches; usually done with lower-grade veneers, allowing for more natural characteristics.

Additionally, hardwood plywood may be specified as **matching architectural panels** in premanufactured sets, sequence-matched uniform size panels, and blueprint matched panels, which ensure grain continuity throughout an entire room or building.

Special effects matching is available in patterns such as checkerboard, diamond, and sunburst.

For more information, contact:

American Plywood Association
P.O. Box 11700
Tacoma, WA 98411-7265

Hardwood Plywood and Veneer Association
P.O. Box 2789
Reston, VA 22090-0789

# APPENDIX 5 | Cedar Shakes and Shingles

Red and white cedar possesses certain qualities that make it suitable for the manufacture of weather-resistant pieced cladding. Cedar **shakes** are, by definition, a split product, while cedar **shingles** are produced by sawing.

Cedar shingles and shakes are produced according to standards set by trade associations, depending on region and species. It is advisable to obtain and review trade association literature prior to the specification of cedar shingles and shakes. An understanding of species, grades within species, regularly stocked products, and local preferences is useful, as each of these factors can affect material cost and availability. Most suppliers stock only popular types; other types must be obtained by special arrangement.

## Shake and Shingle Specification Checklist

- **use**
- **species**
- **grade and rules agency**
- **length**
- **exposure**
- **pattern** (where applicable)

Many of these products are appropriate for both roof and wall applications. Wall applications allow a greater maximum exposure than roof applications, and hence fewer shingles may be needed to cover a given area. Larger exposures require longer shingles. Shingles may be double-coursed to obtain large exposures (12 in. and more) yet these applications must use exposed (butt) nailing to properly restrain the long, exposed ends.

## Grades

Red cedar and white cedar shingles are graded according to the portion (from the butt) of clear wood, as well as other quality-related factors such as the percentage of heartwood and the grain orientation. Red cedar shakes are a specialty product, typically hand-split in a number of configurations. Shakes are usually produced from only the highest-grade wood, although lesser grades are available. Grade affects maximum exposure as well as use. Only the highest grades are suitable for roofs, and large exposures require a higher grade to maintain clear wood exposed to the weather. Indoor decorative applications are less demanding, and any grade may be used for various effects.

Both red cedar and white cedar shingles use a color label system to identify grade. These are listed in Table A5-1. For specific grade and product information, contact the agencies listed.

### Table A5-1  Shingle Grades

| Certigrade Red Cedar Shingles | White Cedar Shingles |
|---|---|
| No.1 Blue Label<br>No. 1 Rebutted and Rejointed[1]<br>No. 1 Machine Grooved | A - Blue Label |
| No. 2 Red Label<br>No. 2 Rebutted & Rejointed[1] | B - Red Label |
| No. 3 Black Label | C - Black Label |
| Undercourse | Utility - Green Label<br>Builder Shims - Yellow |

[1] Machine trimmed with parallel edges and squared butts; for tight-fitting sidewall applications.

For more information, contact:

Cedar Shake and Shingle Bureau
515 116th Ave. NE, Suite 275
Bellevue, WA 98004

Quebec Lumber Manufacturers Association
5055 Hamel Blvd.
Hamel West, Suite 200
Quebec, Quebec G2E-2G6
Canada

# Index

Adhesives, 4, 11
Astragal, 96, 97, 193

Backband, for casing, 89, 100, 194
Balustrades
    deck, 82
    porch, 80
    stair, 149–152
Baseboard, 109–112, 192
Brickmould casing, 30, 44, 192
Building components, elemental, 14–15, 18

Cabinetry, 133–139
    conventions, 134, 145
    face frame, 135, 138, 140–141
    frameless, 135, 136–138
    modular, 134–139
    site built, 139–141
Casing
    door, 98–100
    patterns, 192
    window, 87–89
Caulks, 4, 11, 174–175
Ceilings, 101–108
    beams, 103–104
    concealed lighting, 107
    crown mouldings, 105–108

INDEX

Ceilings *(continued)*
    paneled, 102–103
Chair rail, 109, 194
Closet woodwork, 139–142
    conventions, 135, 142
    details, 141–142
Contemporary details, discussion of, 18–19
Corbel, 132
Corner block, 89
Crown moulding, 105–108, 193
Curves, 155–166
    kerf bending, 163
    long-dimension curves, 156–162
    short-dimension curves, 162–165
    stack lamination, 161
    types of curvature, 156

Decks, 80–84
    balustrade details, 82
    bench details, 82–83
    construction of, 81
    design considerations for, 80–81
Doors, 35–46, 93–100
    choosing, 40–41
    construction of, 36–44
    exterior details, 44–46
    as fields, 35–44
    frame and panel, 37–40

hollow, 94–95
interior details, 93–100
operation, 95–98
pocket, 98, 99
Drip cap, 56, 193
Drip edge
    in roof construction, 62, 65
    ventilated, 68, 69

Exterior protection, 167–176
    applied protection, 173–176
    design protection, 167–172

Facades, 47–59
Fascia, 62, 65–69
Fields and edges, 16–18
Fireplace construction, 126–127
Fireplace woodwork, 125–132
    design considerations, 127–130
    details, 131–132
Flashing, 30–32, 44–45, 57–59, 169–170
Floors, interior, 113–123
    expansion of, 117, 118
    finish options, 115–116
    installation considerations, 118–119

parquet, 120–121
transition details, 121–122
wood strip and plank, 114–120
Frame and panel construction, 37–40
Frieze, 56, 62, 69

Hinges, door, 42–43
Historical precedents, 13–20
Housing, stair, 143

Kerf bending, 163

Lumber, 6–9
  plain-sawn, 7
  quarter-sawn, 7
  vertical-grained, 7

M

Mantel, 128–129, 131–132
Materials, overview, 3–12
Medium density overlay, 206

Metals, 3, 10
Mortise, 38, 40, 95
Mouldings, 191–200
Mullion, 90–92
  factory, 90, 91
  site-built, 90–92

Newel, 149–151

Order, classical, 15–16

P

Paint, 173
Panels, wall, 58–59, 109–112
Parquet flooring, 120–121
Plastics, 4, 11
Plinth block, 100
Plywood, 3, 9–10
  hardwood, 209–210
  softwood, 205–209
Porches, 71–80
  balustrade details, 80
  construction details, 78–80
  construction of, 72, 73

# INDEX

Porches *(continued)*
    design considerations, 72
    floor and foundation details, 73–74
    floor materials, table of, 74
    post details, 75–78
    skirtings, 79

## R

Rails, stair, 149
Relief milling, 114
Reveal, 59
Rise and run of stairs, 148
Risers, stair, 143
Roofline trim, 61–70, 78–79
    brick-veneer detail, 69
    components, 62–63
    design considerations, 63–65
    materials, 64
    of porches, 78–79
    retrofit detail, 66
Rosette, 89

## S

Sealants, 4, 11, 174–175
Sealer, 174
Shakes, cedar, 213–215
Shingles, cedar, 52, 213–215

Shutters, 55–56
Siding, 47–59, 197–204
    corners, 57
    details, 57–59
    materials, 48–50
    patterns, table of, 54
    shakes and shingles, 52, 213–215
    shutters, 55–56
    styles, 51–55
Soffit, 62, 65–69
    open, 66, 68
    ventilation of, 65–69
Spring line, 157
Stain, 174
Stairs, 143–153
    configurations, 146–148
    construction, 143–149
    conventions, 148–149
    rails and balustrades, 149–153
Stool, window, 88
Stringer, stair, 143–146
Sub-fascia, 62

## T

Tenon, 38, 40
Traditional details, discussion of, 18–19
Treads, stair, 143

# Index

Veneer, 210

Wainscot, 109–112, 165–166
Walls, exterior, *see* Siding
    paneled, 58–59
Walls, interior, 109–112
    baseboard, 109–112, 192
    curved wainscot, 165–166
    panels, 109–112
    wainscots, 109–112
Weather-stripping, 43

Window, 23–33, 87–92
    construction of, 24–29
    exterior details, 29–33
    as fields, 23–29
    glass, 25–27
    interior details, 87–92
    multiple, 89–92
    types, 29–33
Wood, 3, 4–10
    characteristics table of, 7
    grades, 201–204
    lumber, 6–9
    structure and properties, 4–6
    types, 6